PENGUIN BOOKS

THE *NEW SCIENTIST* INS

Richard Fifield was managing editor of *New Scientist* from 1969 to 1988 when he became executive editor of the magazine. Previously he was on the editorial staff of *Nature* and *World Medicine*. He has written and edited numerous books on science, medicine and technology, including the *New Scientist* guide *The Making of the Earth*, and the Scientific Committee on Antarctic Research's *International Research in the Antarctic*.

THE *New Scientist* Inside Science

EDITED BY RICHARD FIFIELD

Illustrated by Peter Gardiner

PENGUIN BOOKS

PENGUIN BOOKS

Published by the Penguin Group
27 Wrights Lane, London W8 5TZ, England
Penguin Books USA Inc., 375 Hudson Street, New York, New York 10014, USA
Penguin Books Australia Ltd, Ringwood, Victoria, Australia
Penguin Books Canada Ltd, 10 Alcorn Avenue, Toronto, Ontario, Canada M4V 3B2
Penguin Books (NZ) Ltd, 182–190 Wairau Road, Auckland 10, New Zealand

Penguin Books Ltd, Registered Offices: Harmondsworth, Middlesex, England

This edition first published in Penguin Books 1992
5 7 9 10 8 6 4

Printed in England by Clays Ltd, St Ives plc
Set in Monophoto Times

Contents

Foreword

Towards the end of 1987, *New Scientist* launched a regular supplement that aimed to bring together two ideas – a schools' briefing section that tackled aspects of the science behind the news, and a way of answering some of those everyday questions that most of us want raised but just don't like to ask.

Inside Science set out to tackle such topics as the big bang, embryo research and the ozone holes, and to do it in everyday language and in a form more directly useful to school teachers than the week-to-week features in *New Scientist*.

The supplements soon proved popular with teachers and also with our general readers. The teachers, wishing us well with the project, said they liked the style of the supplements and thanked us not only for not making them just fact and more fact but also for airing some of the wider issues involved. 'They provide examples helping us to bring science alive and to show just how relevant it all is today,' was one teacher's appreciative verdict. The research community also took an interest, and offered to help with refereeing. And we had some unexpected readers. The Cabinet Office, central to policy making in Britain, sent around a large car especially to collect copies of 'The Greenhouse Effect' and 'The Ozone Layer'.

The supplements are the product of science writers, science teachers, research scientists and the magazine's editors. From the beginning, Peter Gardiner has worked tirelessly to illustrate the series. The policy has been to make minimal use of photographs. Here artist, editors and scientist have worked together to produce explanatory diagrams.

On the whole, we have tried to steer away from topics that are handled well in readily available textbooks. Sometimes we have chosen exciting new research, such as oncogenes and the body's protein weapons. At other times we have chosen to tackle older topics that have gained a new relevance, bacterial contamination of food, for example, or complex issues that are relevant to the new science curriculum, such as the structure of the Earth and the life of the stars.

In this, our first selection of the Inside Science supplements, I have brought together those that tackle aspects of the more basic sciences – the physical, astronomical and biological sciences with a strong lacing of chemistry – and have included the first of our mathematical considerations, risk assessment, as it treads across so many topics and issues.

Many people have played a part in the success of these supplements, too many to give everyone a mention. I would, however, like to thank David Dickson, as the editor of *New Scientist*, and Michael Kenward, as the former editor, for their encouragement; John Holman, Andrew Brown and David Sang, all senior science teachers, who tested many of the early manuscripts for us; and my colleagues Jackie Wilson, Omar Sattaur, Mick Hamer, Nina Hall, Stephanie Pain, Peter Wrobel and Jane Anderson who, together with design artists Sarah Reynolds, Rob Morton, Christine Jones and Colin Brewster, have guided the supplements through the press and made this book possible.

Richard Fifield
Executive Editor
New Scientist

Contributors

Frances Balkwill is head of the biological therapy laboratory at the Imperial Cancer Research Fund in London

Marcus Chown is science news editor of *New Scientist*

Heather Couper is an astronomy broadcaster and former president of the British Astronomical Association

Tony Cox is a lecturer in elements inorganic chemistry at the University of Oxford

Georgina Ferry, a former review editor and science news editor of *New Scientist*, is now a press officer for the University of Oxford

Richard Fifield is executive editor of *New Scientist*

Linda Gamlin is a freelance science writer and editor

John Gribbin is physics consultant to *New Scientist*

Nigel Henbest is astronomy consultant to *New Scientist*

Nina Morgan is a science writer specializing in Earth sciences

Fred Pearce is a freelance writer on environmental issues and a former news editor of *New Scientist*

Omar Sattaur is a freelance writer and broadcaster and a former science news editor of *New Scientist*

Ian Stewart is a professor of mathematics at the University of Warwick

Christine Sutton is a freelance writer and a former physical sciences editor of *New Scientist*

Richard Vile is a former research scientist at the Chester Beatty Laboratories, Institute of Cancer Research, London

Ian Woodward is a lecturer in botany at the University of Cambridge

PART 1

Physical and Chemical Sciences

CHAPTER 1

The Big Bang

Marcus Chown

Fifteen thousand million years ago, the Universe that we inhabit erupted, literally, out of nothing. It exploded in a titanic fireball called the big bang. Everything – all matter, energy, even space and time – came into being at that precise instant.

In the earliest moments of the big bang the stuff of the Universe occupied an extraordinarily small volume and was unimaginably hot. It was a seething cauldron of **electromagnetic radiation** mixed with microscopic particles of matter unlike any found in today's Universe. As the fireball expanded it cooled, and more and more structure began to 'freeze out'. Step by step the fundamental particles we know today, building blocks of all ordinary matter, acquired their present identity. The particles condensed into atoms, galaxies began to grow, then fragment into stars such as our Sun. Just over 4000 million years ago, the Earth formed. The rest, as they say, is history.

It is a tremendously grand picture of creation. Yet astronomers and physicists, armed with a growing mass of evidence to back their theories, are so confident of the scenario that they believe they can identify the detailed conditions in the early Universe as it evolved, instant by instant. Unfortunately, when it comes to answering the ultimate question – just how could space and time, matter and energy have come out of absolutely nothing – the theories are not yet good enough. The best that physics can do is to attempt to describe what was happening when the Universe was already about 10^{-35} seconds old – a time that can also be written as a decimal point followed by 34 zeros and a one. This is an exceedingly small interval of time, but you would be wrong if you thought that it was so close to the moment of creation as to make no difference. Although the structure of the Universe no longer changes much in even a million years, when the Universe was young, things changed much more rapidly.

Getting back to the moment of creation

Physicists can run the expansion of the Universe backwards. In this way, they are able to watch it get hotter and hotter as it gets smaller, just as the air in a bicycle pump heats up as it is compressed. But theory predicts that, at the big bang itself, the temperature was infinite. Infinities warn physicists that their theory is flawed.

At the moment, the theories which take us furthest back in time are the Grand Unified Theories. These GUTs are an attempt to show that three of the basic forces that govern the behaviour of all matter are no more than facets of a single superforce.

Each force of nature arises from the exchange of a different 'messenger' particle. The messenger transmits a force between two particles, just as a tennis ball transmits to a player the force of his opponent's shot. At high enough temperatures – such as those that occurred when the Universe was 10^{-35} seconds old – physicists believe the electromagnetic, and strong and weak nuclear forces were identical, and mediated by a messenger dubbed the X-boson.

Physicists want to show that gravity, too, is a facet of the superforce. They suspect that gravity split apart from the other three forces at about 10^{-43} seconds after the big bang. But before they can 'unify' the four forces, they must describe gravity by 'quantum' theory, a type of theory hugely successful in describing the other forces. Physicists are currently finding this difficult.

When they have their unified theory, physicists believe that they will be able to probe right back to the moment of creation and explain how the Universe popped suddenly into existence 15 000 million years ago.

Energy and matter
Two faces of the same coin

Physicists think that as many important events happened between the end of the first tenth of a second and the end of the first second as in the interval from the first hundredth of a second to the first tenth of a second, and so on, logarithmically, back to the very beginning. As they run the history of the Universe back, like a movie in reverse, space is filled with ever more frenzied activity. This is because the early Universe was dominated by electromagnetic radiation – in the form of little packets of energy called **photons**. And the higher the temperature, the more energetic the photons. Now, high-energy photons can change into particles of matter because, as Einstein revealed, energy and matter are simply different faces of the same coin. They are connected by the famous equation $E = mc^2$, where c is the speed of light.

What Einstein's equation says is that particles of a particular mass, m, can be created if the packets of radiation, the photons,

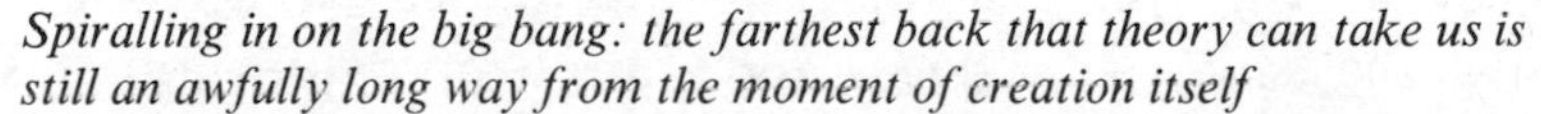

Spiralling in on the big bang: the farthest back that theory can take us is still an awfully long way from the moment of creation itself

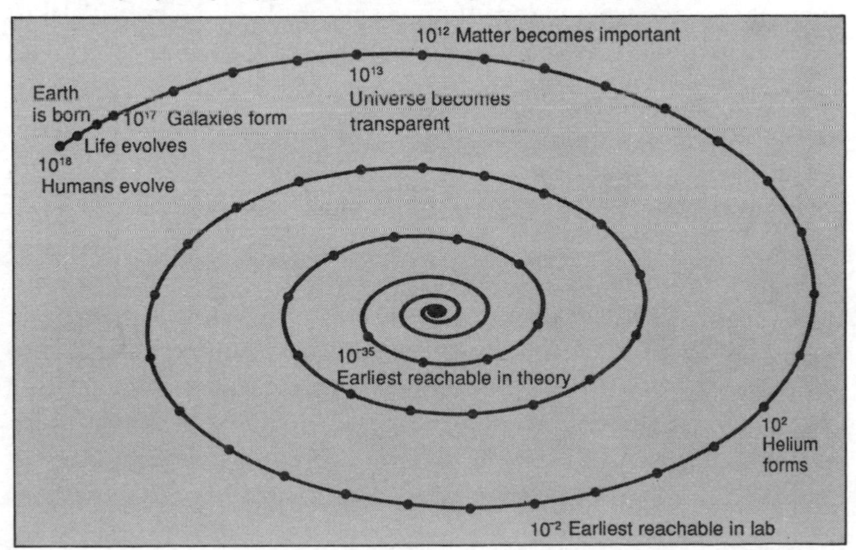

have an energy of at least mc^2. There is, therefore, a temperature above which the photons are energetic enough to produce a particle of mass, m, and below which they cannot create that particle. If we look far enough back, we come to a time when the temperature was so high, and the photons so energetic, that photons could collide and produce particles out of pure energy. What those particles were before the Universe was 10^{-35} seconds old, we do not know. All we can say is that they were very much more massive than the particles we are familiar with today, such as the proton and electron.

As time progressed, and the temperature fell, so the mix of particles in the Universe changed, to a soup of less and less massive particles. Each particle was 'king for a day', or at least for a split second. For the reverse process was also going on – matter was being converted back to energy as particles collided to produce photons.

What do physicists think the Universe was like a mere 10^{-35} seconds after the big bang? Well, the volume of space that was destined to become the entire visible Universe, thousands of millions of light years across, was contained in a volume roughly the size of a pea. And the temperature of this superdense material was an unimaginable 10^{28} degrees Celsius. At this temperature, physicists predict, colliding photons had just the right amount of energy to produce a particle called the **X-boson**. This particle was 1000 million million times more massive than the proton. No one has yet observed an X-boson, because to do so we would have to re-create, in an Earth-bound laboratory, the extreme conditions that existed just 10^{-35} seconds after the big bang.

How far back can physicists probe in their laboratories? The answer is to a time when the Universe was about one hundredth of a second old. At that time, the Universe had grown to fill a volume roughly the size of the Sun. By then, it had cooled down to 10^{14} or 100 million million °C – still many millions of times hotter than the centre of the Sun. But this temperature is not beyond the reach of experiments. In 1983 physicists at

How do we know there was a big bang?

Our modern picture of the Universe is due in large part to an American astronomer, Edwin Hubble. In 1923 he proved that the Milky Way, the great island of stars to which our Sun belongs, was just one galaxy among thousands of millions of others scattered throughout space.

Hubble also found that the wavelength of the light from most of the galaxies was 'red shifted'.

Astronomers interpreted this as a Doppler effect, familiar to anyone who has noticed how the pitch of a police siren changes as it passes by. The siren becomes deeper because the wavelength of the sound is stretched out. Similarly with light, the wavelength of light from a galaxy which is moving away from us is stretched out to the longest, or reddest, wavelength.

Hubble had discovered that most galaxies are receding from the Milky Way. In other words the Universe is expanding. And the farther away the galaxy, the faster it is receding.

One conclusion, therefore, was inescapable: the Universe must have been smaller in the past. There must have been a moment when the Universe started expanding: the moment of its birth. By imagining the expansion running backwards, astronomers deduce that the Universe came into existence about 15 000 million years ago.

This idea of a big bang means that the red shifts of galaxies are not really Doppler shifts. They arise because in the time that light from distant galaxies has

In an expanding universe (top), space continually grows. If such a universe contains enough matter, gravity will make it shrink again and maybe cause it to oscillate (middle). In a steady-state universe, matter is continually created as space expands (bottom)

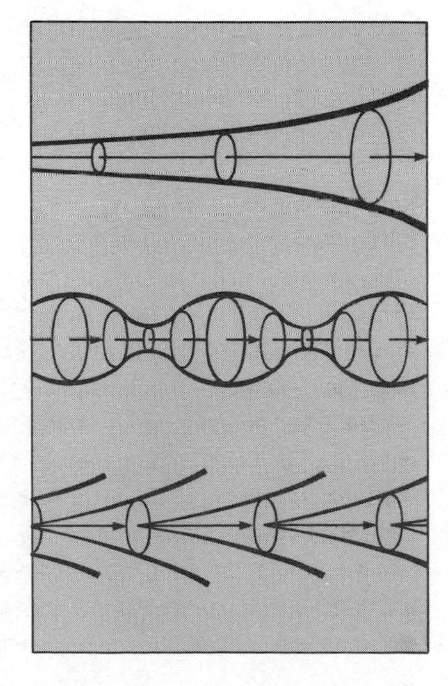

been travelling across space to us, the Universe has grown, stretching the wavelength of light.

The picture of a universe which is expanding need not have been a surprise to anyone. If Albert Einstein had only had faith in his equations, he could have predicted it in 1915 with his theory of gravity – general relativity. But Einstein hung on desperately to the idea that the Universe was static – unchanging, without beginning or end.

The vision of a static universe appealed strongly to astronomers also. In 1948, Hermann Bondi, Thomas Gold and Fred Hoyle proposed the steady-state theory of the Universe. The Universe was expanding, they said, but perhaps it was unchanging in time.

Their theory said that space is expanding at a constant rate but, at the same time, matter is created continuously throughout the Universe. This matter is just enough to compensate for the expansion and keep the density of the Universe constant. Where this matter would come from, nobody could say. But neither could the proponents of the big bang.

The steady-state theory held its own as the principal challenger to the big bang theory for two decades. Then, in the 1960s, two astronomical discoveries dealt it a fatal blow.

The first discovery came from Martin Ryle and his colleagues at Cambridge. They were studying radio galaxies – enormously powerful sources of radio waves, a type of light invisible to the naked eye. In the early 1960s the Cambridge astronomers found that there were many more radio galaxies at large distances than nearby.

The radio waves from these distant objects have taken billions of years to reach us. Ryle and his colleagues, therefore, were observing our universe as it was in an earlier time. The excess of radio galaxies at great distances had to mean that conditions in the remote past were different from those today. A universe which changes with time ran counter to the steady-state theory.

Then, in 1965, Arno Penzias and Robert Wilson, two scientists at the Bell Telephone Labs in New Jersey, detected an odd signal with a radio horn they were using for satellite communications.

The signal did not come from the Earth or the Sun. It seemed to come from all over the sky, and it was equivalent to the energy emitted by a body at three degrees above absolute zero.

There could be no doubt. Penzias and Wilson had discovered the remnant of the radiation from the big bang – the cosmic microwave background. They shared the Nobel prize. The steady-state theory was dead.

CERN (*Centre Européen pour la Recherche Nucléaire*, European Organization for Nuclear Research) in Geneva managed to re-create these conditions in a giant particle accelerator. They created the W- and Z-bosons, particles which vanished from the Universe one hundredth of a second after the big bang.

The gulf between 10^{-35} seconds and one hundredth of a second is gigantic. We know that, for most of this period, matter was squeezed together more tightly than the most compressed matter we know of – that inside the nuclei of atoms. And, as the temperature fell, so the energy level of photons declined, creating smaller and smaller particles. Most were unstable, and collided with each other to produce photons.

Small contaminants
Particles of the future

At some point the hypothetical building blocks of the neutron and proton – known as **quarks** – came into being. Unfortunately, no one has developed a satisfactory theory which explains how a quark soup behaves, so we know little about this period. By about one hundredth of a second, however, the Universe had cooled sufficiently to be dominated by particles that are familiar to us today: photons, electrons, positrons and neutrinos. Neutrons and protons were around, but there weren't many of them. In fact, they were a very small contaminant in the Universe. About one second into the life of the Universe, the temperature had fallen to around 10 000 million °C, and photons had too little energy to produce particles easily.

The next important stage in the history of the Universe was at about 100 seconds. The temperature had dropped to a mere 1000 million °C – the temperature in the heart of the hottest stars. Now the particles were moving more slowly. In the case of the protons and neutrons, this meant that they stayed close to each other long enough for the strong nuclear forces, which bind them together in the nuclei of atoms, to have a chance to take a hold. In particular, two protons and two neutrons could combine to form nuclei of helium. Solitary neutrons decay into protons in about 15 minutes, so any neutrons that were left over after the helium formed became protons. According to physicists' calculations, roughly 10 protons were left over for every helium nucleus that formed. And these became the nuclei of hydrogen atoms, which consist of a single proton.

This is one of the strongest pieces of evidence that the big bang really did happen. For much, much later, when the temperature had cooled considerably, the hydrogen and helium nuclei picked up electrons to become stable atoms. Today, when astronomers measure the abundance of elements in the Universe – in stars and galaxies and interstellar space – they still find roughly one helium atom for every 10 hydrogen atoms.

At one time almost all the electrons and their positively charged opposites, the positrons, were colliding and cancelling each other out by forming photons. There were roughly 1000 million photons for every proton and neutron in the Universe, a ratio which persists to this day. But a slight lopsidedness in the laws of physics meant that at about half an hour after the big bang, at the end of all the collisions, there was a tiny number of electrons remaining. The point at which it was cool enough for these electrons to combine with protons to make the first atoms was a long way down the stream of time – 300 000 years after the big bang. The Universe was now cooling very much more slowly than in its early moments, and the temperature had reached a modest 3000 °C. This also marked another significant event in the early history of the Universe.

The fate of the Universe

The big bang was not at all like the explosion of a lump of material in which fragments are blown away into an existing void. There was no void. Space itself popped suddenly into existence 15 billion years ago and began expanding.

So, when astronomers look at a distant galaxy and find that it appears to be rushing away from us, it is because the *space* between us and the galaxy is swelling in the aftermath of the big bang. Imagine a balloon with dots drawn on its surface to represent galaxies.

Now imagine inflating the balloon. The dots, or galaxies, move apart. They do not move within the surface, they move because the surface expands. The real Universe is a three-dimensional version of the surface of the balloon, and so difficult to imagine.

The flaw in the analogy is that the dots on the balloon grow as the balloon inflates. Galaxies do not grow as the Universe expands – the gravity of each galaxy is strong enough to bind its stars together.

Will the expansion of the Universe continue for ever? This depends on how much matter there is in the Universe. The gravity of each galaxy tries to pull every other galaxy towards it. Ever since the big bang, the gravity of the galaxies has been acting as a brake on the expansion of the Universe.

If there is enough matter in the Universe, gravity will eventually slow, then reverse the expansion until all the matter is recompressed into a tiny volume: a 'big crunch'. The Universe might then rebound in another cycle of big bang and big crunch. If, however, there is not enough matter in the Universe, the expansion will continue.

The galaxies are rushing apart because the space between them is stretching in the aftermath of the big bang

Until the electrons had combined with the hydrogen and helium nuclei, photons could not travel far in a straight line without running into an electron. Free electrons are very good at scattering, or redirecting, photons. As a consequence, every photon had to zigzag its way across the Universe. This had the effect of making the Universe opaque. If, for instance, the light from the stars zigzagged its way across space to your eyes rather than flew in straight lines, on a clear night you would see only a dim milky glow across the whole sky rather than a myriad of stars. We can still detect photons from this period. No longer creating matter, they have been flying freely through the Universe for about 15 000 million years, and astronomers observe them as the so-called **cosmic microwave background**. Whereas these photons started their journey when the temperature was 3000 °C, the Universe has expanded 1000

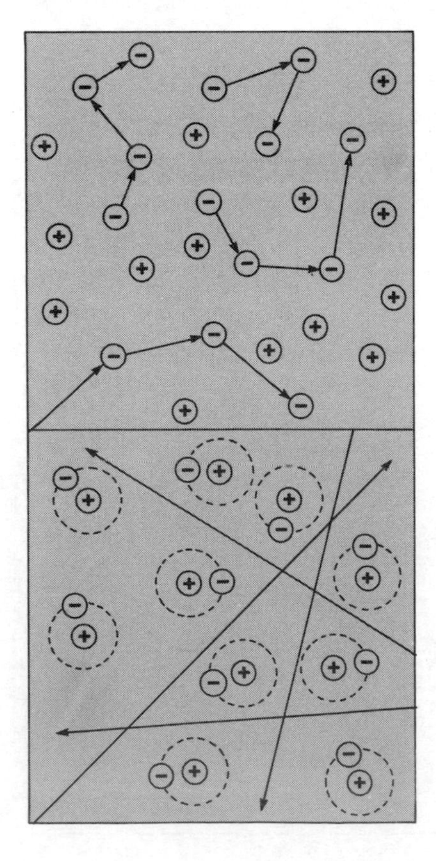

When it was cool enough for electrons and protons to form atoms, the Universe became transparent. Photons, which had been scattered by free electrons (top), could now fly unhindered (bottom)

times while they have been in flight. This has decreased their energy by this factor, so that we now record the signals as just 3 °C above absolute zero. The temperature dropping to about 3000 °C also signalled another event – the point at which the energy levels of the radiation, or photons, in the Universe fell below that of the matter. From then on, the Universe was dominated by matter and by the forces of gravity acting on that matter.

The building of elements, however, had stopped abruptly after the Universe had reached an age of 100 seconds and the protons and neutrons had formed the nuclei of hydrogen and helium. For elements such as carbon and oxygen to form, higher temperatures were needed, but the Universe was getting colder all the while. The heavy elements in the planets and in your body were created, billions of years later, in the nuclear furnaces of stars (see Chapter 7).

As the Universe continued to expand, gravity caused clumps of matter to accumulate in large islands. Those islands were to become the galaxies. The galaxies continued their headlong rush into the void, fragmenting into smaller clumps which became individual stars, producing heat and light by nuclear reactions deep in their core. At one point, about 10 000 million years after the big bang, a yellow star was born towards the outer edge of a great spiral whirlpool of stars called the Milky Way.

The star was our Sun.

22 October 1987

Further Reading

The First Three Minutes by Steven Weinberg (Fontana, 1983) is the best popular account of the standard model of the big bang. Because it was written 13 years ago, however, it does not describe the first hundredth of a second. *Superforce* by Paul Davies (Heinemann, 1984) describes the quest to unify the forces of nature. It may be heavy going, though, for anyone without a science background. *In Search of the Big Bang* by John Gribbin (Corgi, 1987) is a popular account of the big bang which speculates on the precise nature of the big bang and what happened before.

CHAPTER 2

Four Fundamental Forces

Christine Sutton

One of the more remarkable discoveries of the past century is that the great diversity of our Universe stems from a handful of essential building blocks – the subatomic particles. Less well advertised, but equally remarkable, has been the discovery that these particles interact in a few basic ways. Physicists speak in terms of the **fundamental forces** that take part in these interactions between particles and mould the Universe into the form we observe. The use of the term 'force' in this sense may be confusing. We often come across forces in the more restricted mechanical sense – the force required to push open a door, for example. Physicists today, however, know that all the familiar forces result from the underlying fundamental forces.

In mechanics a force is something that can change the state of motion of an object, as when a golf club hits a stationary ball. Or, forces can maintain the position of an object, as when someone holds a ladder steady against a wall. The early Greek philosopher Aristotle, who lived in the fourth century BC, believed that motion always requires a force. In other words, some force or other must continue to act on a golf ball to keep it moving after the impact from the club. It was not until nearly 2000 years later that the Italian scientist Galileo Galilei recognized that it is not motion itself that requires a force, but changes in motion. Galileo realized that if there were no forces to act upon a golf ball, for example, then once hit it would continue for ever in a straight line through space, at a constant velocity. This is not, of course, what happens. Air resistance tends to slow down the ball and the wind can alter its direction. Moreover, gravity ensures that the ball, initially directed upwards, comes heading back to the ground on a sweeping arc.

Henry Cavendish with the famous torsion balance experiment that determined the gravitational constant G and demonstrated Newton's inverse square law of gravitation. Large lead spheres placed close to small ones caused angular deflection

Galileo's work was soon consolidated by the English physicist and mathematician Isaac Newton. In his famous *Principia Mathematica*, published in 1687, Newton wrote down his three famous laws of motion. Even today these laws tell us all we need to know about mechanical forces. Newton's contribution to mechanics is honoured in the name given to the SI unit of force – the newton. This is defined as the force needed to accelerate an object with a mass of one kilogram by one metre per second per second. In more practical terms, this is about equivalent to the force required (on Earth) to lift a small apple.

With the aid of his three laws, Newton dealt successfully with the force of gravity. He recognized that gravity is not only a property of the Earth – which allows apocryphal apples to fall from trees – but a property of all objects with mass. In other words, the same force keeps the Moon in its orbit around the Earth as causes apples to fall to the ground. Indeed, gravity holds whole galaxies together.

Matter of attraction
Inverse square law

According to Newton's law of gravitation, the force of attraction, F, between any two objects is proportional to the product of their two masses, m_1 and m_2, divided by the square of the distance, r, between them:

$$F = \frac{Gm_1m_2}{r^2}$$

The intrinsic strength of the **gravitational force** is, however, the same in all cases. It is given by the constant of proportionality in Newton's law of gravitation – the **gravitational constant**, symbolized by G.

As far as physicists can tell, the gravitational constant G seems to be a universal constant. That is, it is the same in all parts of the Universe, although some research has indicated that its value may have changed during the history of the Universe. The value of G is currently accepted as 6.672×10^{-11} cubic metres per kilogram per second per second ($m^3kg^{-1}s^{-2}$).

The gravitational force is today recognized as being the weakest of the fundamental forces. It is a force that derives from the existence of **mass**. Physicists say that mass is the **source** of the force of gravity; putting it another way, mass is the source of a **field** that maps the strength and direction of the gravitational force at all points in space (and time).

The concept of a field of force was formulated in the nineteenth century in studies of the forces in electricity and

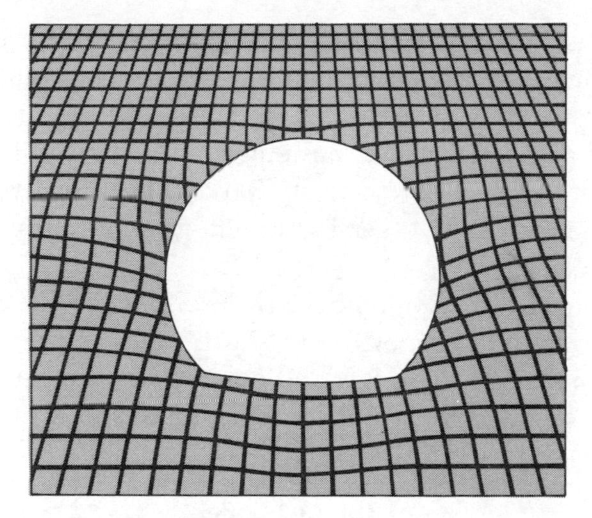

The gravitational field around a large object like the Sun

magnetism. Try to bring the north poles of two magnets together and you will find that it is very difficult. There is a repulsive force between the two similar poles. Or rub a balloon on your sweater or jumper and then hold the balloon up to the ceiling. The balloon will stay there, held by an attractive electrical force. In 1785 the French physicist Charles Coulomb worked out in detail how the force between electrically charged objects obeys a rule similar to Newton's law of gravitation. Both forces follow an **inverse square law**. In other words, the strength of the force is inversely proportional to the square of the distance between the objects involved. With electrically charged objects, the magnitude of the force, F, is proportional to the product of the sizes of the two charges, q_1 and q_2, divided by r^2, the square of the distance between them:

$$F = \frac{q_1 q_2}{\varepsilon r^2}$$

This is Coulomb's law. The constant of proportionality, ε, is a property of the space between the charges.

There is one subtle difference between the gravitational force and the electric force. Electric charge – the source of electrical force – exists in two varieties, known as positive and negative.

By contrast, there is only one type of mass. In all cases, the gravitational force between masses is attractive, pulling them towards each other. The electric force can likewise be attractive, but it can also be repulsive, pushing charged objects away from each other. Charges of the same kind (both positive or both negative) repel each other, while charges of different kinds attract each other.

A similar law to Coulomb's law describes the force between magnetic charges, or north and south poles. Again the force obeys an inverse square law and again the force can be either attractive or repulsive: like poles repel, unlike poles attract.

Lines of force
Electromagnetic fields

One person to make major contributions to our understanding of electric and magnetic forces was the English scientist Michael Faraday. He invented the idea of **lines of force**. Imagine that you have an electrically charged object (positive or negative) hanging in the centre of a room. Then suppose that you have

Lines of force depict electric fields (between charges) and magnetic fields (between poles). The lines are shown here in the three-dimensional form in which they extend through space; note that the lines extend outward into space, but in practice have a limited range

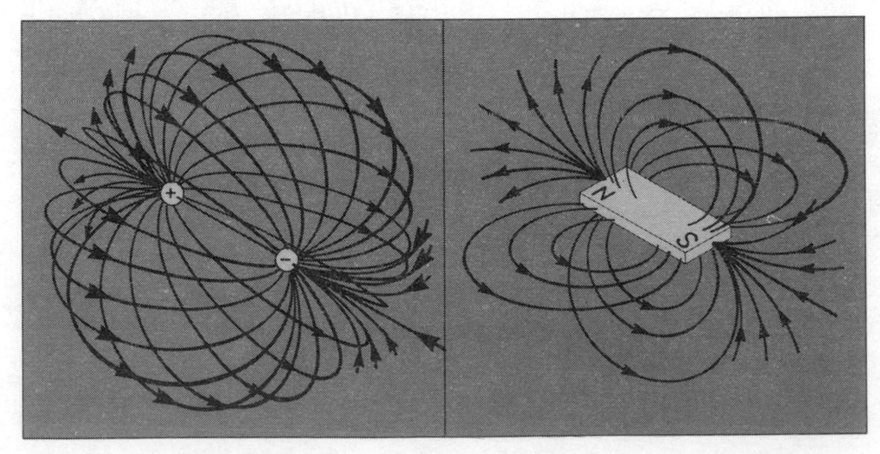

another object with one unit of the same type of charge, and that you can measure the force between the two objects at any point in the room. The direction of the force will always be along the straight line joining the two objects, so you could map the force as lines radiating from the central object – rather like the spines on a sea urchin.

Faraday's lines of force are a valuable aid in visualizing forces. They show the field of a force due to an object or between objects. Indeed we have become accustomed to seeing electric and magnetic fields, in particular, depicted in this way. The field around a magnet becomes still more of a physical reality if we use iron filings and allow them to line up in the direction of the magnetic forces.

Physicists can also represent these fields mathematically. Coulomb's law, for example, yields the force on a unit of charge at any point in space around a charged object. In this way the law allows the field to be calculated. The equation that underlies this calculation at all points in space and time is an important characterization of the force. Attempting to discover this basic field equation (or set of equations) has become a crucial part of understanding the fundamental forces.

The laws describing separate electric and magnetic forces due to stationary, or static, electric and magnetic charges are not, however, fundamental. They are part of a larger picture. Electric and magnetic forces are inextricably linked. For example, moving electric charges (that is, electric currents) produce magnetic effects, while moving magnets can induce electric currents.

All the experimental laws of electricity and magnetism, such as Coulomb's law, come together in the single theory of **electromagnetism**. This great synthesis was performed in the mid nineteenth century by the Scottish physicist James Clerk Maxwell. Maxwell's distillation of the various effects resulted in a set of four basic equations, describing the behaviour of an overall electromagnetic field. An unexpected bonus of Maxwell's work was that his theory turned out to be a theory of light as well as electricity and magnetism. Visible light is an electromagnetic wave with a frequency in a particular range. Similar waves at higher frequencies form the radiation known as X-rays and gamma rays; at lower frequencies, electromagnetic waves are known as microwaves and radiowaves.

Maxwell's equations describe the field of the **electromagnetic force**, a force that ranks alongside gravity as one of the fundamental forces of the Universe. This is the force that holds atoms together, and thereby underlies many of the physical properties of matter in bulk, such as melting points and boiling points, compressibility and elasticity. It is also responsible for chemical reactions, allowing atoms to combine in different

Four forces

There are four forces which between them control the behaviour of everything in the Universe. Gravity is the weakest force, but it has an infinite range, and every particle in the Universe feels its influence. The electromagnetic force is stronger than gravity; it too has an infinite range.

Electromagnetism is the force that holds atoms and molecules together. Gravity can be ignored on the atomic scale because it is so weak, but its influence increases with every atom that is added to an object, and becomes important once objects contain enough atoms to be visible.

Two other forces, which are described in Chapter 3, control the behaviour of particles on the subatomic scale. They are both very short-range forces, and exert no influence over distances bigger than the diameter of an atomic nucleus, even though one of them is stronger at short range than gravity or electromagnetism.

Because one force is stronger than the other, they are called the **strong** and **weak** nuclear forces.

Physicists sometimes refer to a 'fifth force', which, some experiments have suggested, weakens the influence of gravity slightly over distances of a few score metres. The force, if it exists, is actually a modification to the inverse square law of gravity, and is not an independent force in its own right.

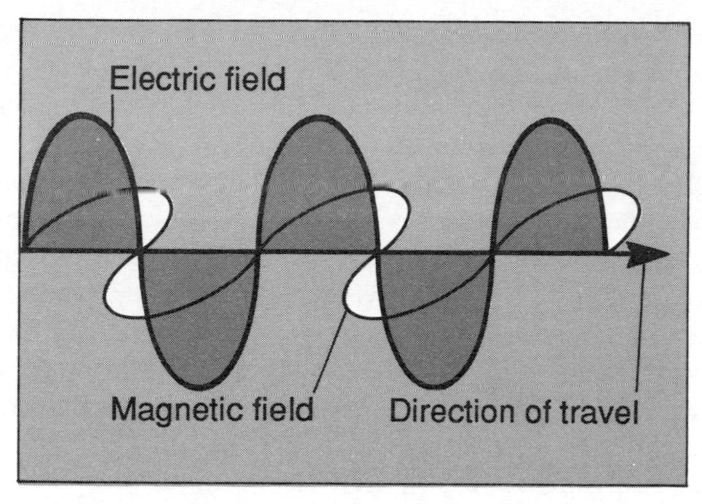

Maxwell's theory of electromagnetism revealed that light consists of electric and magnetic fields vibrating in unison in directions at right angles to each other. The exact wavelength (distance between peaks) varies through a complete spectrum

Attraction between atoms binds matter together; repulsion prevents matter from condensing into a very dense state

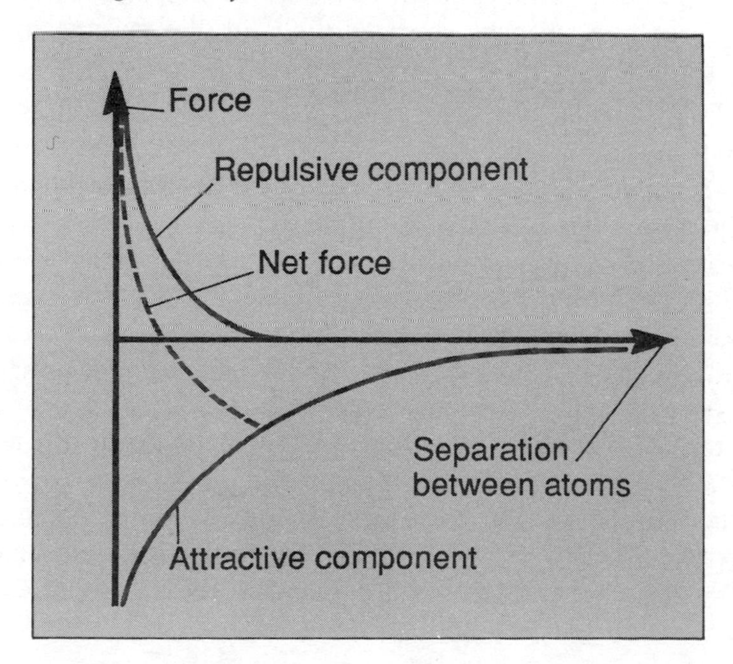

ways to form a huge variety of molecules. In this way the electromagnetic force underlies the complex biochemistry that leads to life itself. An understanding of the electromagnetic force is therefore fundamental not only to physics but also to chemistry and biochemistry. So how does the electromagnetic force hold matter together?

A typical atom consists of negatively charged electrons moving around a tiny, positively charged central nucleus. The attraction between electrical opposites generally keeps the electrons bound to the nucleus. The total negative charge of the electrons exactly balances the positive charge of the nucleus, so that the atom is electrically neutral overall.

Electrical forces are also responsible for holding atoms together, despite the fact that individual atoms have no charge. The binding between atoms depends on an **interatomic force** that has both attractive and repulsive components. The force must be repulsive at the shortest distances, otherwise all matter would condense to a very dense solid state. Beyond these very short distances the force must be attractive to allow atoms to join together in solids and liquids. But the range of this attractive force must also be short enough to allow atoms to drift apart so that materials can exist as gases.

The balance between the repulsive and attractive components of the interatomic force varies from one material to another. Many materials are solids at normal room temperatures here on Earth, while other substances remain as gases even at relatively low temperatures.

The binding between atoms can occur in various ways, but in all cases the force involved is basically electrical – in other words, it arises from the fundamental electromagnetic force. In **ionic compounds**, for example, atoms of one element take electrons from atoms of another element to give both atoms more favourable electronic configurations. But this leaves one element in the form of positive ions (too few electrons) and the other as negative ions (too many electrons). The result is that the oppositely charged ions are attracted to each other electrically, forming a tightly bound material. Ionic compounds such

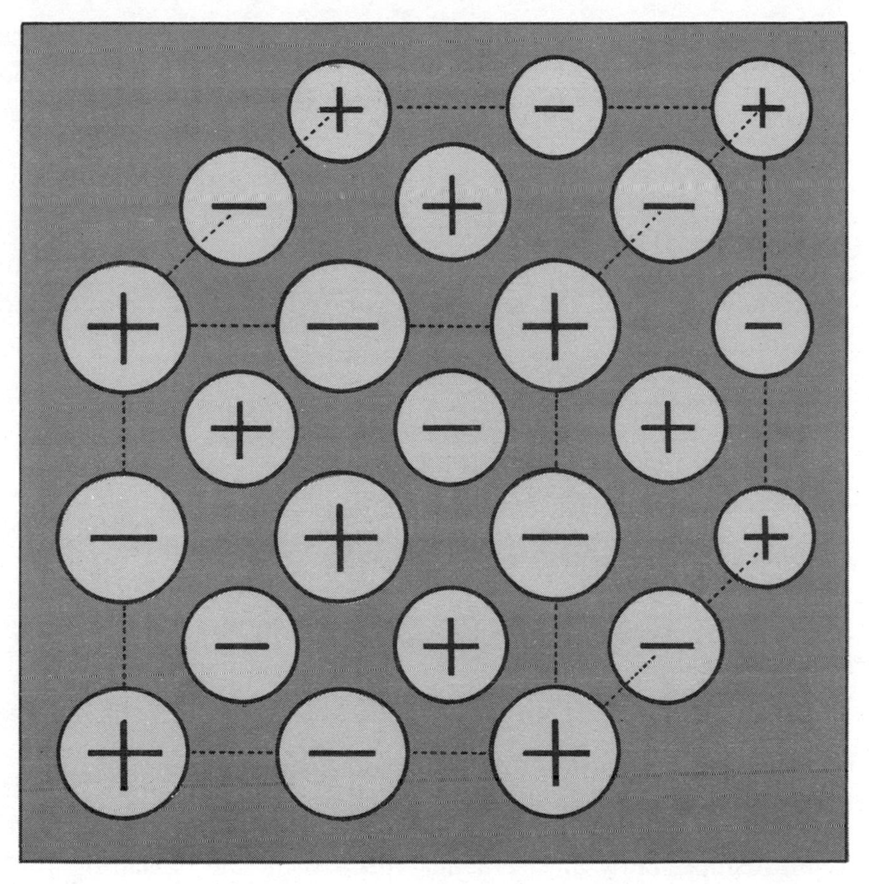

In ionic compounds, such as common salt, atoms borrow electrons from neighbours to achieve better electronic configurations. Now called ions, these have opposite charges and attract each other, binding the compound

as sodium chloride (common salt) have high boiling points and melting points. Indeed, all are solids at room temperature. Such is the influence of the particular type of electrical bonding at work in these materials.

The forces between atoms and molecules are responsible on a larger scale for bulk forces, such as friction. Even the smoothest of surfaces is uneven on the atomic scale, and when two surfaces come together there are tiny regions that are the first to touch. The resulting pressure becomes very great at these points, and the forces between the atoms in the two surfaces give rise to what we call friction.

Subatomic forces
Quantum mechanics

Maxwell's laws of electromagnetism are valuable in describing many processes in the physical world. They are not sufficient, however, to explain in detail the reactions between atoms and molecules, nor even the structure of an individual atom and the way that it absorbs or emits light. To work at the subatomic level, a theory of charged particles must incorporate two specific approaches. In atoms, electrons are moving at high velocities, close to the speed of light. This means that Albert Einstein's special theory of relativity is applicable. Secondly, the appropriate mechanics to apply are not the mechanics of Newton and Galileo, which describe large-scale phenomena. The atom is, instead, the realm of **quantum mechanics**, developed in the 1920s by such people as Niels Bohr, Werner Heisenberg and Erwin Schrödinger.

Briefly, quantum theory deals with atoms in which the energies of the electrons are **quantized** – in other words, the electrons can take up only certain allowed energies. These restrictions lead to a detailed understanding of atomic structure and the relationships between elements that form the basis of the periodic table. A complete theory of any fundamental force must therefore incorporate special relativity and quantum mechanics if it is to be valid universally. With the electromagnetic force, the theory that achieves this is known as **quantum electrodynamics**, or **QED**. Perfected in the 1940s, this theory has been checked against experiment to a high degree of accuracy.

Quantum electrodynamics is an example of a quantum field theory. In QED, electric charge is still the source of an electromagnetic field, as in Maxwell's theory. The difference is that in QED the field has a graininess: it is quantized. According to QED, if we could look closely enough we would see that the field around a charge – around an electron, say – is due to tiny particles flitting in and out like bees around a hive. The field becomes weaker further from the charge because the particles become more spread out.

What are these electromagnetic field particles? The answer is that they are **photons**, the particles or quanta of light. As early as 1900 Max Planck showed that when materials emit or absorb electromagnetic radiation, such as visible light, they emit or absorb a stream of discrete packets of energy – in other words, photons. In quantum theory, electromagnetic waves are seen as photons, the energy of each photon being proportional to the frequency of the radiation. In the same way, photons are the carrier particles of the electromagnetic field. However, these photons have only a brief existence as they flit in and out of an electrical charge. Physicists call these electromagnetic field particles **'virtual' photons**. They do not exist independently in the way that the real photons of electromagnetic radiation do. The theory of QED treats the interactions between all

A particle, for example a photon, 'carries' force from one object to another. The effect here is a repulsive force (the objects move away from each other)

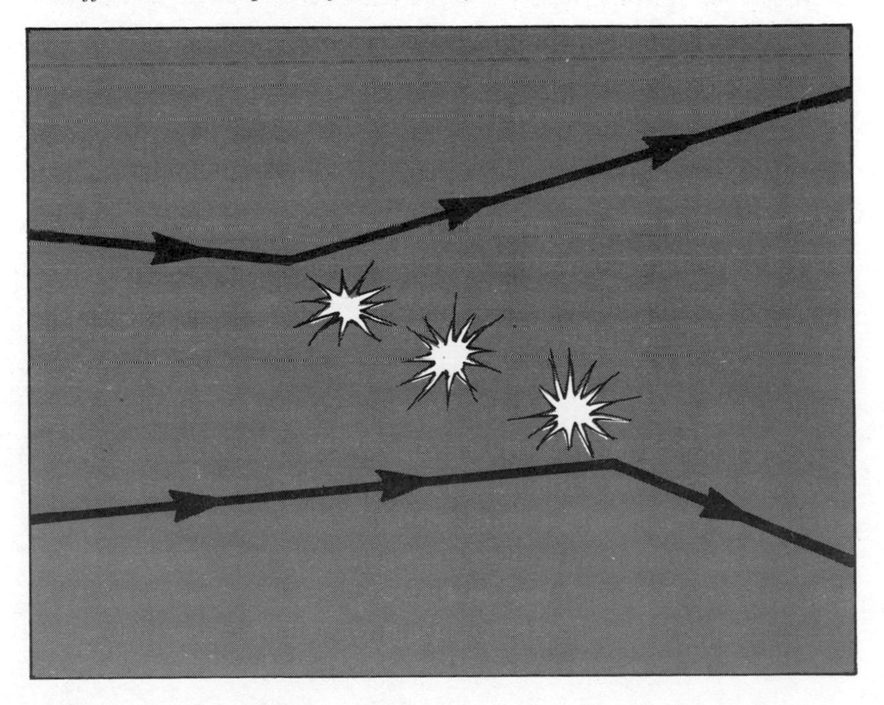

electrically charged particles in terms of the exchange of virtual photons. Two protons repel each other as a virtual photon emitted from one is absorbed by the other. In this way, each charged particle responds to the electromagnetic field of the other.

The strength of this fundamental reaction – the intrinsic strength of the electromagnetic force – is given by a constant known as the fine structure constant, symbolized by *a*. This constant, which takes its name from a detailed effect in atomic spectra, is a pure number, with no units. Its value is 1/137.036. A comparable pure number, related to the gravitational force between two protons, turns out to be about 5×10^{-39}. This illustrates how gravity is very much weaker than the electromagnetic force.

Problems of gravity
Unified field theory

Does gravity have carrier particles, akin to photons? Physicists do indeed postulate the existence of such particles, which they call **gravitons**. However, a quantum field theory of gravity, similar to QED, has so far eluded theorists.

An important theory of gravity already exists in the form of Albert Einstein's general theory of relativity, which extends special relativity to deal with accelerating masses. This theory does not so much replace Newtonian gravity as incorporate it in a proper field theory on a par with Maxwell's electromagnetic theory. The problems come when physicists attempt to incorporate quantum mechanics with the field theory of general relativity. Calculations begin to go wrong and yield infinite, nonsensical answers. Theorists believe that the solution may lie not in quantizing gravity on its own, but in developing a quantum field theory that incorporates gravity and electromagnetism together with other fundamental forces.

This chapter has shown how physicists can describe much of the world about us with only two fundamental forces – gravity and the electromagnetic force. But this is not the end of the story. Quantum electrodynamics is not enough to explain all the behaviour of matter on the subatomic scale. Chapter 3 describes the additional fundamental forces that work within the atom. It also shows how physicists aim to unite all the fundamental forces within a single comprehensive theory.

19 November 1988

Further Reading

The Hutchinson Encyclopaedia of Science in Everyday Life (Hutchinson, 1988) is a highly illustrated guide to the physical sciences which contains several chapters dealing with gravity, the electromagnetic force and the bonding between atoms and molecules. *Feynman Lectures on Physics*, Volume I, by Richard Feynman, Robert Leighton and Matthew Sands (Addison-Wesley, 1989), contains a more advanced but exceptionally well-explained introduction to the concept of force and its application in physics, especially in Chapter 12. Feynman's ideas were elaborated in his book *The Character of Physical Law* (BBC Publications, 1965). *An Introduction to Concepts and Theories in Physical Science* (second edition), by Gerald Holton, revised by Stephen Brush (Princeton University Press, 1985), includes interesting discussions about force, from a more historical perspective.

CHAPTER 3

Subatomic Forces

Christine Sutton

Atoms, the basic building blocks of life, are mainly empty. If the nucleus of an atom were a centimetre across, then the electron cloud that surrounds the nucleus would be a kilometre away. So why, when you bang your fist against a brick wall, does it hurt? The answer is 'because of the electromagnetic force' – one of the four fundamental forces of nature that we met in Chapter 2.

These four forces are all that physicists need to explain the workings of the everyday world, the Universe at large, and the inside of atoms. Gravity is the most familiar force, and holds matter together in the form of planets, stars and galaxies. The electromagnetic force binds matter on the much smaller atomic scale. It holds the electrons in place around the nucleus, and it holds the atoms themselves in place alongside their neighbours. It is electromagnetic interactions between atoms within a brick that prevent its compression when you hit it, and transmit the impact back into your hand.

The electromagnetic force is not enough, however, to explain all the behaviour of matter on the subatomic scale. Two additional forces operate deep within the atom. Moreover, although these forces are not apparent at larger scales, they have played a part in establishing the Universe as it is today.

Finding forces
Strong attraction

How do we know that these forces exist? One clue comes from the basic structure of matter. **Atoms** consist of a cloud of negatively charged **electrons** that surround a central positively charged **nucleus**, which generally contains two kinds of particle – positive **protons** and neutral **neutrons**.

Each proton carries one unit of positive charge, each electron one unit of negative charge. In an atom, the nucleus contains the same number of protons as there are electrons in the cloud outside, making the atom neutral overall. Opposite charges attract, but only up to a certain point; the electrons cannot 'fall in' to the nucleus because of quantum effects, so they provide the visible face of the atom, largely responsible for its chemical properties. But like charges repel one another, so why does the electric force between protons, packed together in a volume 10^{-14} of a metre across, not blow the nucleus apart? The answer in part lies with the neutral neutrons which 'dilute' the electric force between protons. But then, how are the neutrons held there?

There must be another fundamental force operating within nuclei, stronger than the electric force. This holds the protons (and neutrons) together within the nucleus. But this force has no influence outside the nucleus, so it must, unlike gravity or electromagnetism, be limited to a very short range – only about 10^{-15} metres. It is called the **strong force**, and experiments involving collisions between fundamental particles show that within the nucleus it is about 100 times stronger than the electric force between protons.

Colourful quarks
Fundamental truth

Where does the strong force come from? To answer this, physicists have tried to develop an understanding of the strong force using the same kind of description that works so well for electromagnetism. The electromagnetic force is the one that physicists understand best, and which is described by a very satisfactory theory known as quantum electrodynamics, or

QED (see Chapter 2). This uses quantum field theory – it describes the 'field' about an object in terms of continual absorption and emission of fundamental particles, known as field quanta.

In QED, the object is anything that carries electric charge, the field is the electromagnetic field, and the field quanta are **photons** – 'particles' of light. An electric charge is constantly emitting and absorbing individual **'virtual' photons**. If another charged object absorbs one of these photons, then the objects have influenced each other – they have interacted via the electromagnetic force. By analogy with QED, physicists interpret the strong force in terms of a property that is equivalent to electric charge, associated with field quanta that are equivalent to photons. This understanding arose after experiments revealed a deeper level to the structure of matter.

These experiments showed that protons and neutrons are composed of more fundamental particles, called **quarks**. Quarks, like electrons, seem to be truly fundamental particles that cannot be divided further. They carry electric charges which are either one third or two thirds the size of the standard unit of charge on an electron or proton. But they come in combinations that ensure that the charge always adds up to whole units or zero. For example, there are three quarks inside every proton and every neutron. The property of quarks that is like electric charge but involves the strong force is called their **colour charge**. This name arose because it appears to come in three varieties, like the primary colours of light; but it has nothing to do with colour in the everyday sense of the word. It is just that it is easy to think in familiar concrete terms, so the three types of charge associated with the strong force are given the names red, green and blue.

Rather as positive and negative electric charge can add up to zero, so the colour charges of quarks can add up to give no colour. This is what happens in protons and neutrons. These particles are 'colourless', even though they contain coloured quarks – the proton, for example, contains one blue quark, one red and one green. In a similar way an atom is 'chargeless',

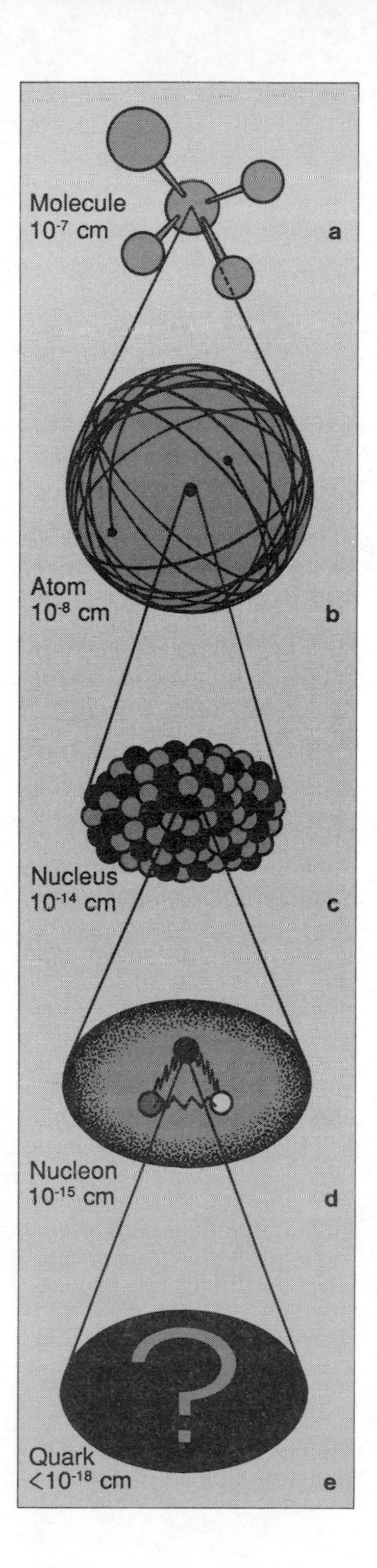

The electromagnetic force holds matter together on atomic and molecular scales. But it is the strong force that holds the constituents of atomic nuclei together, and binds the quarks within individual nucleons (a) Residual effects of the electromagnetic force bind electrically neutral atoms together in molecules. (b) The electromagnetic force holds the atom together, binding a 'cloud' of electrons around a central nucleus. (c) Residual effects of the strong force bind nucleons together in nuclei. (d) The strong force holds quarks together in the nucleons. (e) No experiment so far has revealed any structure within a quark.

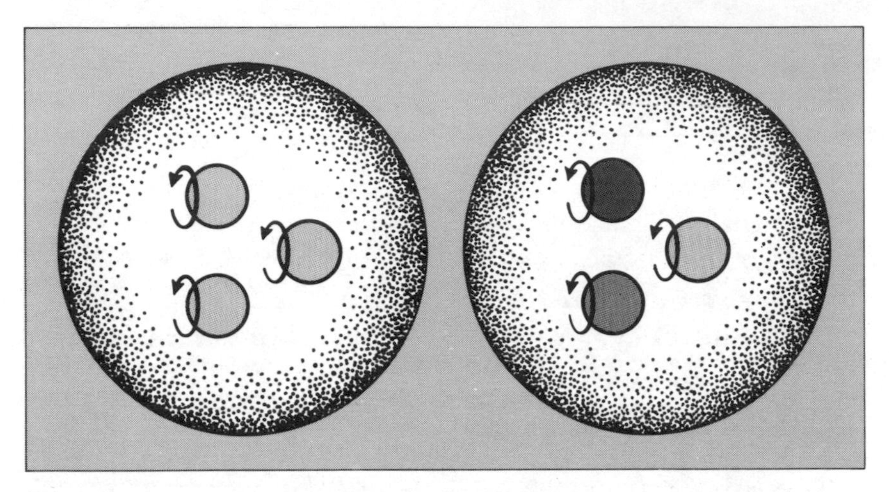

The particle called the delta-double-plus presented physicists with a problem. According to the quark model it contains three quarks of the same kind, all spinning the same way. But this violates a quantum rule, known as Pauli's exclusion principle, which prohibits two particles from being in an identical state. The problem is resolved if quarks can exist with three kinds of colour charge. Then, the quarks in the delta-double-plus are distinct from one another and are not subject to the exclusion rule

even though it contains both positive protons and negative electrons.

So the strong force, which affects only coloured particles, does not operate directly on protons or neutrons. It operates on the quarks within them. But it can still hold protons and neutrons together in the nucleus, in the same way that electromagnetic forces can hold electrically neutral atoms together. The positively charged nucleus of an atom is only partly screened by its own electron cloud, and feels the influence of negatively charged electrons in the cloud surrounding the neighbouring atom, giving rise to Van der Waals forces; similarly, the coloured quarks inside a proton feel the presence of quarks in the proton next door.

Nuclear glue
Ties that bind

Electromagnetic forces involve the exchange of photons; the equivalent field particles that carry the strong force are called **gluons**, because they 'glue' particles together. Field lines depict the forces between quarks, in the same way that they are used to depict forces between electrically charged particles. Because this whole theory is based on QED, but involves so-called colour charges, it is known as **quantum chromodynamics**, or **QCD**.

Like photons, gluons have no mass. But there is one particularly important difference between gluons and photons. Photons are not electrically charged, and so do not interact with each other through the electromagnetic force. Gluons, however, carry colour charge, as quarks do – there are eight different kinds of gluon, each with its own combination of colours. So gluons can interact with each other as well as with

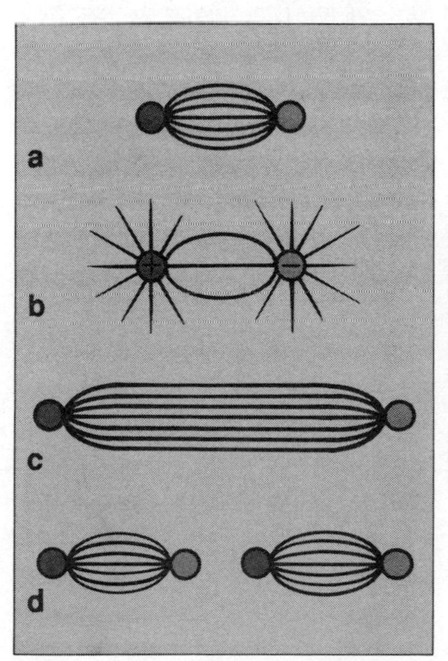

The field lines of the strong force between quarks form a tightly defined region in space (a), unlike the electric field between charged particles (b). As the quarks move apart, the strong field lines remain parallel (c), indicating that the force remains as strong. Eventually, new quarks emerge from the energy of the field (d), so forming a new particle. (For simplicity, the diagrams illustrate particles containing a quark and antiquark, with opposite colour charges. Particles made from two quarks do not actually exist because their colour charges do not neutralize)

quarks via the strong force. This makes that force very different from the electromagnetic force. When two electrically charged particles are pulled apart, the force between them decreases. But this does not happen with colour charge. The force between two coloured quarks does not decrease as the distance between them increases. Instead, the gluons associated with the quarks pull on each other as well as on the quarks. This seems to be the reason why a lone quark can never escape from inside a proton, but must always exist in a colourless combination with other quarks. However, no one has yet proved theoretically that quarks and gluons must always be confined in this way.

Even the strong force, however, is not sufficient to explain all the behaviour of particles in nuclei and outside them. For example, a neutron that is free from the confines of an atomic nucleus does not live for ever. After typically a little under 15 minutes, an isolated neutron will spit out an electron and a

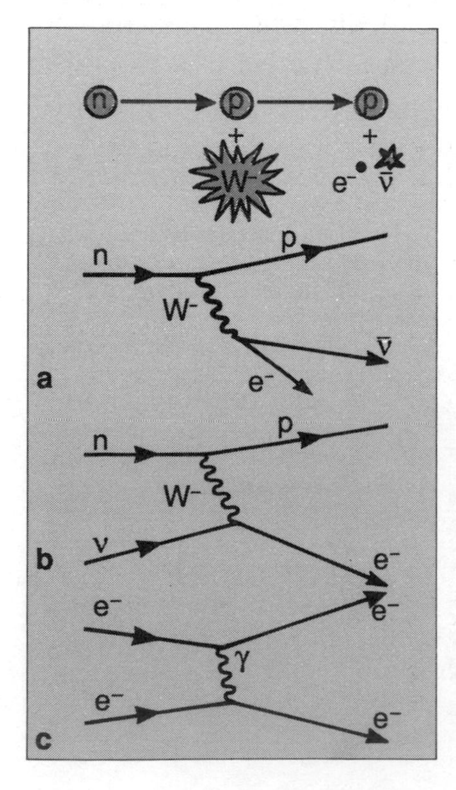

When a neutron (n) decays, it converts into a proton (p) by emitting a W^- particle, as in **a**. *The W^- almost immediately converts into an electron (e^-) and an antineutrino ($\bar{\nu}$). The W^- is one of the carriers of the weak force. The outgoing antineutrino, $\bar{\nu}$, can be replaced by an incoming neutrino, ν, without changing the physics, as in* ***b***. *This reaction via the weak force then looks very similar to the reaction between charged particles due to the electromagnetic force,* **c**, *which occurs through the exchange of a photon (γ)*

particle called an **antineutrino**, and will become a proton. A neutron inside a nucleus, however, does not do this, except in unstable radioactive nuclei. Physicists can describe the 'decay' of the neutron in terms of the emission and absorption of field quanta, and therefore in terms of a fundamental force. In this case, the force is called the **weak force**, because within the nucleus it is about 10 000 times weaker than the strong force.

There are three field particles that transmit the weak force. Two of these are electrically charged. These are called W^+ and W^-. The third force carrier, which is uncharged, is known as the Z°. So the weak force may change the charge of a particle, as when a neutron decays into a proton, or it may involve interactions in which there is no charge change.

All three weak force carriers are heavy – each has a mass roughly 100 times that of a proton or a neutron. So how can a neutron emit a virtual W or Z particle that is much heavier than itself, even if that particle is later reabsorbed? The answer lies in **quantum uncertainty**. Quantum physics shows us that no quantity, such as the mass of a neutron, can ever be fixed precisely. There is always some uncertainty in the amount of mass – strictly speaking, mass-energy – associated with a particle, or even with a point in empty space. Over a long period of time, this uncertainty is very small. But over a very short period of time, the uncertainty is very large. A neutron can create, and emit, a W or Z particle out of nothing at all, provided that the particle is absorbed by the neutron or another particle in such a short time that the Universe does not notice the discrepancy. The time allowed depends on the mass of the particle (longer if it is lighter) and is fixed by the quantum rules.

Photons, the field quanta of the electromagnetic force, have zero rest mass. These particles can travel for ever across the Universe. But W and Z particles are heavy, and so cannot travel far from their parents. So, like the strong force, but for a different reason, the weak force has a very short range. Physicists have been able to create W and Z particles by colliding beams of subatomic particles together at such high energies that the amount of mass available, in line with $E = mc^2$, exceeds the mass of these carriers of the weak force. Then, the particles are real, not virtual. Their existence can be traced through detectors such as those at the European research centre CERN and their masses can be determined. Such experiments help us to investigate the role of forces in the early Universe. When the Universe was very young, just after the big bang, it was hot and dense (see Chapter 1). Space was filled with energetic radiation and particles of all kinds. This energetic radiation is what we now detect as a weak hiss of radio noise at a temperature of 3 °C above absolute zero, coming from all directions in space – the cosmic background radiation. In the first moments of the existence of the Universe, this radiation was at a temperature of billions of degrees, and under those conditions W and Z particles could be made as easily as photons.

Two into one
Electroweak force

Theory suggests, and high-energy experiments confirm, that at high enough energies electromagnetism and the weak force are parts of a single, unified **electroweak force**. They only split into two forces at lower energies – lower temperatures. Physicists discovered the connection between the weak and electromagnetic forces not through experiments but in their attempt to develop a quantum field theory for the weak force. Indirect confirmation of the electroweak theory came first in the early 1970s; but it was only in 1984 that researchers at CERN succeeded in observing the direct production of W and Z particles. This proved two forces could be combined in one package.

This observation underlies an important discovery. One way to develop a feel for how two forces can be combined into one is to recall that electricity and magnetism seem at first sight to be unrelated phenomena, yet they are manifestations of the same underlying electromagnetic force. A totally unrelated analogy may help. The 'dripping' from roast meat appears as a

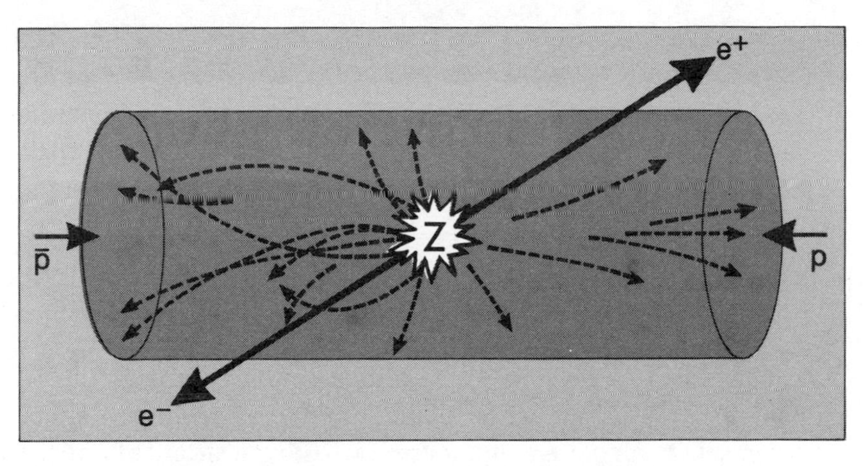

A decaying Z particle betrays its presence in the straight back-to-back tracks due to an electron–positron pair ($e^- e^+$), at CERN. The tracks are picked up in the gas-filled central detector, which observes collisions of high-energy protons (p) and antiprotons ($\bar{p}$)

homogeneous liquid when it is hot, straight from the oven. But as it cools down to normal room temperature it separates into two distinct components – a fatty, more or less white solid on top and a brown jelly below. We inhabit a world where the two components of the electroweak force have separated out, like dripping, to give the appearance of two distinct forces.

Into the future
Ultimate physics

The success of unifying two of the four forces in one theory has encouraged physicists to search for a single 'theory of everything' describing the electroweak, strong and gravitational forces as facets of a single underlying force. Such a theory might answer such questions as 'what is mass?' and would certainly provide insights into conditions in the early stages of the Universe, when it was so hot that all four forces were indistinguishable.

Towards a theory of everything

Because of the fundamental importance of the basic forces, physicists have an intense interest in discovering if a **fully unified theory**, describing all four forces with one set of equations, does exist. Such a theory has some great difficulties to overcome. In particular, it must encompass a quantum field theory of gravity.

The standard theory of gravity is Einstein's theory, general relativity. The **theory of relativity** links space and time, and within this theory what we are used to thinking of as the force of gravity is described in terms of the distortion of space-time caused by the presence of matter. But the space and time of general relativity are both smooth and continuous. In this sense, general relativity is a **'classical' theory**. It does not incorporate the essential idea from quantum physics that everything comes in discrete units, or quanta.

A physicist trained only in Newtonian theory would find it easy to understand general relativity, but much harder to understand quantum physics. The other three fundamental forces are all describable in quantum terms, and this raises hope that all three can be incorporated into a unified theory. But gravity must first be quantized before it can be included in the package.

Even the individual field theories have run into problems. In particular, they were plagued by infinities which emerged from the equations and could only be removed by a trick known as renormalization. In effect, renormalization amounts to dividing both sides of an equation by infinity, to 'cancel out' the unwanted terms.

This is not a very satisfactory state of affairs since, strictly speaking, dividing one infinity by another could give any answer. But at least it could be made to work. When gravity was included in the package, however, the trick could not be made to work. Infinities still emerged from the equations, but these infinities could not be cancelled. Quantum gravity seemed to be non-renormalizable.

A breakthrough came when some theorists developed a description of fundamental particles not as mathematical points but as tiny one-dimensional entities, or **strings**. This changed the structure of the equations describing interactions between particles in two important ways. First, the equations now include, automatically, the description of a quantum field particle that has exactly the properties required to be the carrier of the gravitational force – the **graviton**. Secondly, in some variations of these equations all the infinities disappear from the equations automatically, with no need for renormalization. The infinities only disappear if gravity is included; gravity can only be made to fit if the infinities disappear.

This powerful discovery suggests to many physicists that the new ideas, which go by the name of **string theory**, are a step on the road to a true theory of everything. The strings themselves are tiny. It would take 10^{20} of them (a 1 followed by 20 zeros) laid end to end to stretch across an atomic nucleus. But because they are 'spread out', even by such a small amount, they require a fundamentally different set of equations for their description. It will take many years for the implications to be worked out and the complete theory, if one exists, to be found. But then, in a sense, science will have achieved its ultimate goal.

That may seem far removed from the practicalities of our existence today. Yet the present differences between the four forces are crucial to the state of the world about us. The Sun, for example, is fuelled by nuclear reactions that have their basis in the weak force, but is held together by gravity. If the weak force were a little stronger, compared with gravity, the Sun would have burned out long ago; if it were weaker, the Sun's output of energy would be more feeble. Either way, there might then be no life on Earth to ponder the nature of the four forces.

If a unified theory can be developed, it will tell us more about each of the four component forces, and therefore about the nature of the world we live in today.

11 February 1989

Further Reading

Feynman Lectures on Physics by Richard Feynman and others (Addison-Wesley, 1989). *Building the Universe* edited by Christine Sutton (Basil Blackwell, 1985) provides an overview of forces and particles in the form of articles culled from the pages of *New Scientist*. *Superforce* by Paul Davies (Heinemann, 1984) and *Superstrings: a Theory of Everything?* edited by Paul Davies and Julian Brown (Cambridge University Press, 1988) are both readable accounts of the search for a unified theory. *In Search of Schrödinger's Cat* by John Gribbin (Corgi, 1984) explains the underlying quantum principles. *The Particle Connection* by Christine Sutton (Hutchinson, 1984) goes into a little more detail.

CHAPTER 4

Quantum Rules, OK!

John Gribbin

Quantum physics is a *practical* theory with many applications that impinge on our daily lives. It underlies recent developments in computers, telecommunications and genetic engineering. And yet, at the heart of quantum physics there is a set of concepts so bizarre that it is impossible to understand them in everyday terms. Quantum mechanics deals with chance and uncertainty, a world where particles are waves and waves are particles, and the act of observing a system changes the state it is in. The quantum mystery is, indeed, so mysterious that one of the quantum pioneers, the American physicist Richard Feynman, once said that '*nobody* understands it'. But at least it is possible to present the mystery, in all its glory, for you to puzzle over.

At the beginning of the twentieth century the German physicist Max Planck realized that the nature of the spectrum of radiation emitted by a hot object – the so-called **black body curve** – could be explained if radiation was considered to be made up of small units, or packets, just as matter is made up of atoms. He called these units **quanta**.

The radiation from a hot object always has a characteristic black body form. There is very little radiation at long wavelengths and very little at short wavelengths, with a peak of intensity somewhere in between. This peak shifts to shorter wavelengths as the body gets hotter (from infrared, to red, to blue, to ultraviolet, and so on). 'Classical' wave theory, treating the electromagnetic waves in the same way that you would treat vibrations of a violin string, told nineteenth-century physicists that there should always be a huge amount of radiation emitted at very short wavelengths – that is, in the

The ideal absorber and radiator

A black body is the physicist's ideal absorber of energy – 'black' because it absorbs all electromagnetic radiation that falls upon it. But because the laws that describe the behaviour of radiation can be reversed in time, such a perfect *absorber* is also a perfect radiator.

A hot black body is the most efficient radiator of electromagnetic energy allowed by the laws of physics. The name has stuck, even though, to a physicist, a 'black body' might be red-hot or even white-hot!

A white-hot object radiates with peak intensity in the optical spectrum, so our eyes see a mixture of wavelengths corresponding to all colours of the rainbow as white light.

The spectrum of solar radiation is essentially the black body curve for an object at 6000 °C

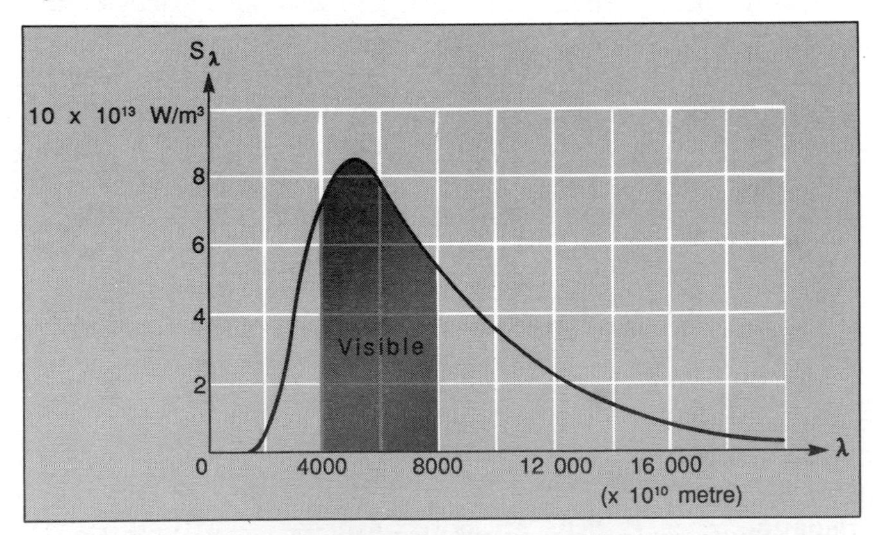

ultraviolet region – but there just isn't. According to classical theory the energy emitted in any waveband is proportional to the frequency and inversely proportional to the wavelength. So, as the wavelength goes to zero, the energy tends to infinity. This puzzle became known as the **Ultraviolet catastrophe**.

The catastrophe was resolved by Planck's suggestion that radiation – such as light – could be emitted only in packets greater than a certain size. Instead of the available energy being spread out continuously (in effect, made up of an infinite number of tiny pieces), as classical theory required, Planck's new theory described the statistics of electromagnetic energy divided up into a *finite* number of pieces. That statistical description exactly matched the observed black body curve. It said that the energy E of each piece of radiation is related to its frequency f by the equation

$$E = hf$$

where h is a constant now known as Planck's constant.

It is easy to see how this resolves the ultraviolet catastrophe. For very high frequencies the energy that is needed to emit one quantum of radiation is large, because f is large, and only a few emitters will possess that much energy. At very low frequencies there are many electrons with enough energy to emit the appropriate low-energy quanta, but these each carry so little energy that even added together they don't amount to much. Only in the middle of the black body curve, where there are a moderate number of electrons each with enough energy to emit moderately large quanta, is there a peak of radiation.

The detailed statistics explain the observations perfectly. But since every physicist believed, early in the twentieth century, that electromagnetic radiation was a wave phenomenon, quantization was initially thought to be something to do with the structure of atoms, rather than the nature of electromagnetic waves. Atoms were thought to emit only certain quantities of radiation, even though 'in-between' quantities existed, rather like the way in which an automatic cash dispenser at the bank will issue you with money only in multiples of £5, even though other values (such as £17.87) exist in the world outside.

Even in this early form, quantum theory helped the development, by the early 1920s, of an understanding of the structure of atoms. An atom consists of a positively charged nucleus, surrounded by a cloud of negatively charged electrons. Since

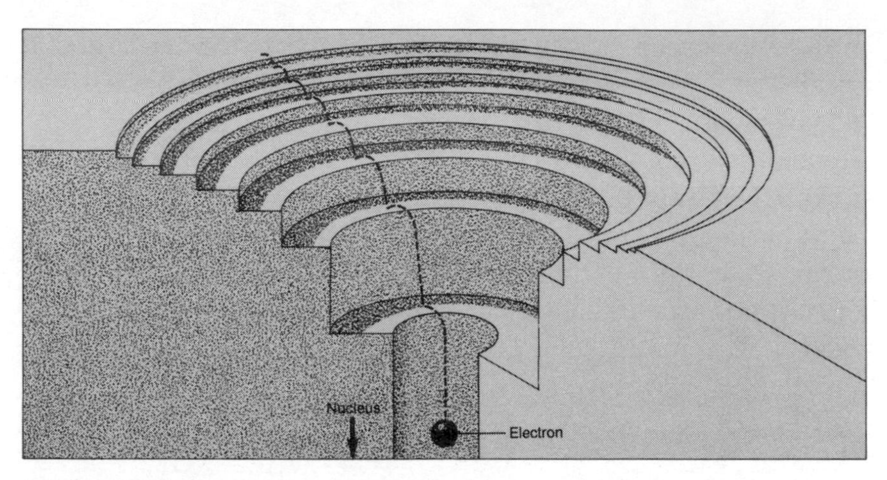

Energy levels in a simple atom such as hydrogen can be compared with a set of steps having different depths. A ball placed on different steps represents an electron in different energy levels. Moving down from one step to another releases a precise amount of energy. 'In-between' amounts of radiation cannot be emitted because there is no 'in-between' step for the electron to land on

opposite charges attract, why don't the negative electrons fall in to the positive nucleus?

If they did so, they would radiate energy continuously as they fell. According to quantum theory the electrons can occupy

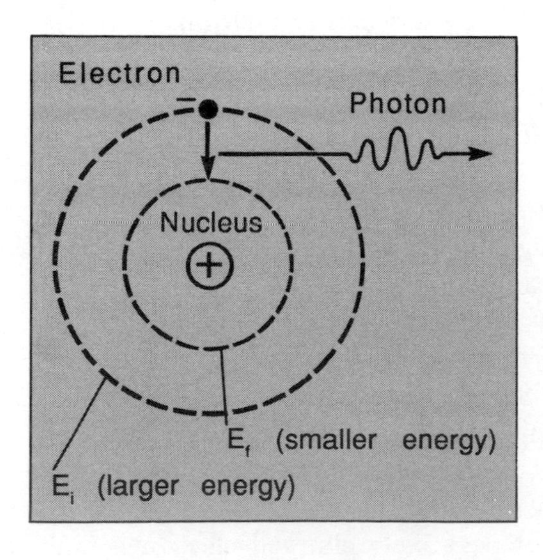

A jumping electron emits a photon

only certain well-defined **energy levels** around the nucleus. An electron can jump from one energy level to another, emitting or absorbing the appropriate quantum of energy as it does so. But it cannot jump to any in-between state, because there are no in-between states. Electrons are kept apart from one another, with a limited number allowed on each energy level, and cannot fall into the nucleus.

All this could be explained even if the energy itself – light or other electromagnetic radiation – came with any amount of energy, rather than being quantized. But Albert Einstein had made a more dramatic proposal. In 1905 he pointed out that the way in which electrons are knocked out of a metal surface by radiation (the photoelectric effect) could be explained only if light itself was 'quantized' – in the form of a stream of particles, now known as photons. The properties required of photons to explain the photoelectric effect were exactly the properties required to explain the black body curve, with electromagnetic energy only *allowed* to exist in discrete lumps, as if money did not exist except in multiples of £5 – as if you could have £5, £10, £15, but *not* £12.37. (The size of each 'lump' of light depends on the wavelength – but for a specific wavelength the energy carried by each photon is the same.)

This was the work for which Einstein eventually received the Nobel prize for physics in 1921. But it threw physics into confusion because there was already a wealth of evidence that light – electromagnetic radiation – was a wave phenomenon. In particular, by shining a beam of light through two narrow slits, the Englishman Thomas Young and the French physicist Augustin Fresnel had each shown, early in the nineteenth century, that light waves produce an **interference pattern** just like the interference produced by two sets of ripples moving across the surface of a pond. By the 1920s the double-slit experiment *proved* that light was a wave, while the photoelectric effect *proved* that light was a stream of particles. Worse was to come.

Particles, waves and the experiment with two holes

Japanese researchers have recently carried out the definitive version of the experiment with two holes. In the standard double-slit experiment for light, a beam of light from a single source is passed through two slits in a screen, and on to another screen. Light waves travelling to a point on the second screen by the two routes travel a different number of wavelengths before reaching it. Where the waves are in step, they add together to produce a bright stripe on the screen; where they are out of step, they cancel to leave a dark stripe. The stripes are **interference fringes**.

The Japanese team did the same thing by shooting electrons, one at a time, through an instrument known as an electron biprism, and monitoring the build-up of spots of light caused by the arrival of the electrons at a TV screen (*a* to *c* in the diagram). As the image builds up, it forms dark and light stripes – interference

When two waves meet, they interfere with each other. Thomas Young used this early in the nineteenth century to prove that light is a form of waves. Light from a source passes through two slits in a screen to produce two sets of waves. Where the waves cancel out, they leave dark shadows on a second screen; where they add together, they produce bright bands of light. Exactly equivalent interference fringes are produced when electrons are fired one at a time through a two-slit apparatus

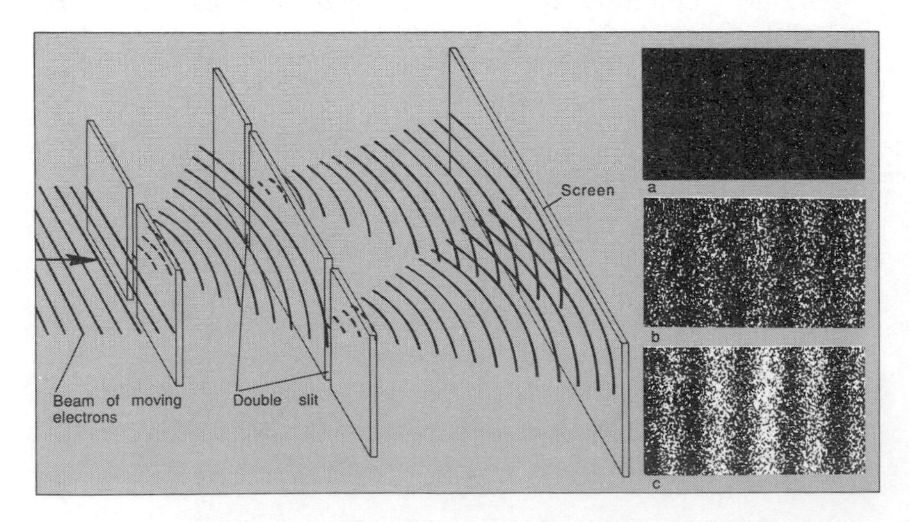

fringes. Each electron behaves like a particle when it strikes the screen, arriving at one spot. But the position it arrives at seems to have been determined by a wave moving through *both* slits of the apparatus. It is a wave and a particle at the same time.

The mystery deepens
Electron waves

In 1897 the English physicist J. J. Thomson (always referred to by his initials) found negatively charged particles with a mass $\frac{1}{1837}$ that of the hydrogen atom. He had discovered the electron and identified it as a component of the atom. He received the Nobel prize for physics in 1906, for identifying the electron as a particle in its own right. In the 1920s, however, the Frenchman Louis de Broglie adapted Einstein's photon description of light to electrons. The equations which showed light 'waves' behaving as 'particles' could be turned around to describe electron 'particles' behaving as 'waves'. Each electron, said de Broglie, had a wave in some sense associated with it, guiding its motion. The permitted energy levels for an electron in an atom corresponded to orbits in which a precise number of wavelengths fitted around the nucleus.

Later in the same decade researchers began to study the way the atoms in a crystal lattice deflected beams of electrons. Among several researchers who carried out such studies (essentially a variation on the familiar double-slit experiment) was George Thomson, the son of J. J. These studies showed that electrons are, under the right conditions, diffracted by the crystal lattice and produce interference patterns. The experiments proved that electrons are waves, and George Thomson shared the Nobel prize for physics in 1937 with the American Clinton Davisson. J. J., who had received a Nobel prize for proving that electrons are *particles*, thus had the satisfaction of seeing his son receive the prize for proving that electrons are *waves*. Both awards were fully merited; father and son were both correct. Electrons behave like particles; electrons behave like waves. Every particle which possesses a momentum p also has a wavelength, λ, and the two are related by de Broglie's equation

$$p = h/\lambda$$

where h is, once again, Planck's constant. The dual nature of particles and waves is only apparent at the atomic and subatomic level, not at the human level, because h is so small – 6.63×10^{-34} joule seconds. The duality is important for electrons, because they have a comparably small mass – just over 9×10^{-31} kilograms.

This **wave–particle duality** is the central mystery of the quantum world. It is closely related to the concept of quantum uncertainty – that we can *never* know both the position and the momentum of a 'particle' with absolute precision at the same time.

In the everyday world a wave is a spread-out thing. The ripples on a pond spread over a long distance, and it is hard to tell exactly where the string of ripples – the wave train – begins and ends. But a particle is a very well-defined thing, which occupies a definite place at a definite time. How can these two conflicting images be reconciled, as they must be if an electron (or a photon) is to be regarded as both wave and particle at the same time?

The appropriate image is of a little package of waves, a short wave train which extends over only a small distance, roughly corresponding to the size of the equivalent particle. Such **wave packets** are easy to describe mathematically. But the way to create a wave packet which is localized in space is to allow many waves of different wavelength to interfere with one another. The smaller the wave packet, the greater the variety of waves with different wavelengths needed to keep it tightly confined.

This spread of wavelengths corresponds to a spread in momentum, since each unique wavelength has its own specific momentum associated with it, in line with de Broglie's equation.

So the *more* precisely the position of the wave packet (= particle) is defined, the *less* precisely its momentum is specified.

Quantum uncertainty
Einstein's pet hate

You can either know where a particle is, or where it is going, but not both at the same time. If we measure the momentum of an electron, say, precisely, then in a sense we are releasing it from the wave packet and selecting a single wavelength for it. That single wave with a pure frequency extends, in principle, to infinity, so the electron then has *no* unique position. But if we measure its position, then we force it into a multi-wavelength state with an uncertain momentum. The very nature of reality depends, at this level, on the kind of measurements we make.

Some people mistakenly think that quantum uncertainty simply indicates the practical difficulty of measuring small things like electrons. Even today, uncertainty is still sometimes described (incorrectly) in terms of the way such measurements might be carried out. In order to observe an electron, the argument runs, we would have to bounce radiation off it, and the very act of prodding it in this way will change its position

A particle can be described in terms of a wave packet, like this one. The packet is spread over a distance X. This distance represents the uncertainty in the position of the particle

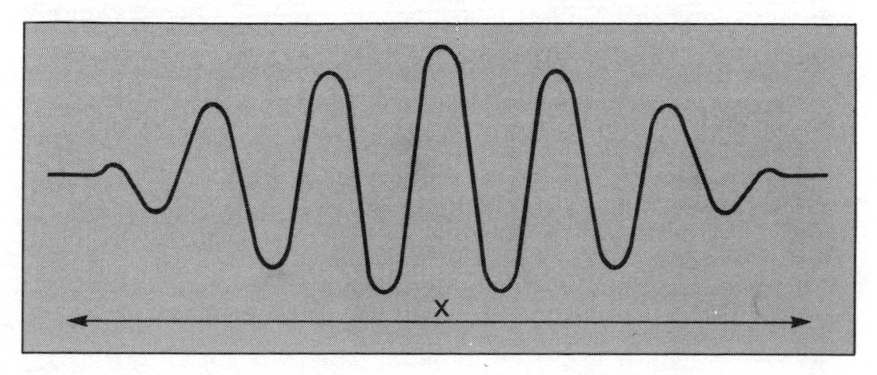

and momentum. That is true, but misses the point. Werner Heisenberg, the German physicist who first appreciated the importance of quantum uncertainty, showed that uncertainty is a fundamental feature of the nature of an electron or other 'particle'. In the quantum world, objects *do not possess* separate properties known as momentum and position; they carry a mixture of the two, a mixture which can never be completely unravelled *in principle*, not just because of experimental limitations. Momentum and position, and the very idea of a particle, are derived from our experience of the macroscopic world. They simply do not work on the microscopic scale.

All of this led the Dane, Niels Bohr (building on the work of the German Max Born), to develop, in the late 1920s, what is still the standard 'explanation' of the quantum world. It is called the **Copenhagen interpretation**, in his honour, and the key features can best be understood in terms of what happens when a scientist makes an experimental observation. First, we have to accept that the very act of observing a thing changes it. We are part of any quantum experiment, and there is no clockwork that ticks away behind the scenes in the same way whether we look or not. Secondly, all we can ever know are the results of experiments. We can look at an electron and find it in position A; then we look again and find it in position B. We guess that it moved from A to B, but we can say nothing at all about how it did so.

What we learn from experiments is that there is a definite **probability** that if we look once and get answer A, then the next time we look we will get answer B (and a corresponding, different probability for answers C, D, E ...). Because there are so many electrons (for example) in everyday systems such as a TV set, the probabilities can be applied with great confidence. Out of each million electrons tweaked in a certain way by an electromagnetic field, a definite proportion, in line with the probabilities, will head off in a certain direction. As long as enough electrons, in a predictable way, travel to the right spot on the TV screen, we don't care how they got there, or what happens to the proportion that end up somewhere else. But

Using quanta

Quantum physics seems like an exotic byway of science to many people, a bizarre theory that has little to do with everyday realities. But that is a misconception. Developments built directly on the quantum theory are a key feature of our everyday lives in the late twentieth century. Curiously, more people have at least a vague understanding of relativity theory, although this has scarcely any impact on our daily lives.

It is quantum theory, for example, that explains the processes both of nuclear fusion and nuclear fission. Einstein's famous equation $E = mc^2$ tells us that mass *might* be converted into energy, but it is quantum physics that tells us *how* to do the trick.

Electrons can be made to 'jump' from one level of energy to another in an atom, and this is a major component of quantum theory. It is also the key to the development of lasers, where electrons sitting on the equivalent energy step in many different atoms are all persuaded to make the same downward jump together, each releasing a photon of the same pure frequency, so that all the photons combine as a laser beam.

A compact disc player depends on quantum physics for its operation. Apart from the laser that scans the disc, quantum effects are crucial in determining the behaviour of electrons in its solid-state circuitry – and in the semiconducting chips of the microcomputer on which I am writing these words.

The quantum even enters the story of life itself. DNA, the famous 'double helix' molecule that carries the genetic code, is held together as a double helix by a chemical effect known as hydrogen bonding, a phenomenon which is explained in terms of the wave nature of single electrons. Unlike a pure particle, even a single electron can spread around an atom, effectively distributing its charge through a cloud. The way this distribution of charge interacts with the positive charge on hydrogen nuclei helps to form the hydrogen bond.

Nuclear power, lasers, computers and the secret of life. Just a few of the things explained, or made possible, by 'esoteric' quantum theory.

underlying such practical considerations, quantum physics deals *only* with probabilities, not with certainties – the discovery that led Einstein, in disgust, to disown it, commenting 'I cannot believe that God plays dice'. The bizarre features of quantum physics that Einstein found so objectionable are clearly seen not in the workings of a TV set but in cases where the probabilities are more evenly distributed.

Einstein didn't believe it; even Richard Feynman never claimed to understand it. But every test that we have been able to apply tells us that at the subatomic level particles and waves are two aspects of a single reality, that the outcome of any interaction depends on chance, and that the way we measure things determines the answers we get. There is no clockwork inexorably guiding the workings of the Universe from the big bang to the end of time.

16 September 1989

Further Reading

Volume III of the classic *Feynman Lectures on Physics* by Richard Feynman, Robert Leighton and Matthew Sands (Addison-Wesley, 1989) is the best textbook guide to the experiment with two holes. Two of John Gribbin's books, *In Search of Schrödinger's Cat* (Corgi, 1985) and *In Search of the Double Helix* (Corgi, 1985) go into more detail about the mysteries of quantum reality and the practical applications of quantum physics. Tony Hey and Patrick Walters tell the story with the aid of lavish illustrations in *The Quantum Universe* (Cambridge University Press, 1987), and Nick Herbert's *Quantum Reality* (Rider, 1985) goes into even deeper mysteries but is still readable.

CHAPTER 5

A Theory of Some Gravity

John Gribbin

Albert Einstein used to say that the unique flash of insight that set him on the path to general relativity came when he realized that a man falling from a roof – or a person trapped inside a freely falling lift – does not feel the force of gravity. People in the falling lift will float, completely weightless, able to push themselves from wall to wall or floor to ceiling with great ease. Of course, we have now seen people in exactly this situation – astronauts in spacecraft, falling freely in orbit around the Earth. In such 'weightless' conditions objects obey, precisely, Newton's famous laws of motion, proceeding in straight lines unless interfered with by forces. But Einstein had to imagine all the things we have seen for ourselves on television – pencils hanging in mid air, liquids that refuse to pour, and so on. Einstein's genius saw all this, and the important point missed by everybody else. If the acceleration of the falling lift, plunging downwards at an ever increasing speed, can *precisely* cancel out the force of gravity, then that force and acceleration are exactly equivalent to each other.

The power of this insight – the **principle of equivalence** – is clear if we imagine the lift replaced by a closed laboratory which is being accelerated through space by a constant force. Everything in the laboratory falls to the floor, and a physicist who carries out experiments inside will be unable to tell whether the downward force is due to an acceleration or to the force of gravity pulling things down. We are used to thinking of acceleration as being caused by a force, but from the point of view of the lift's occupant, that force is caused by acceleration.

Now, said Einstein, imagine setting up an experiment inside

Gravity and uniform acceleration produce identical forces, which we call weight. No experiment can distinguish between them

such a laboratory to measure the behaviour of a beam of light that crosses from one side of the room to the other. In a laboratory moving at constant velocity, far from any planet or star, the light will travel in a straight line across the laboratory. But in an accelerated laboratory, the opposite wall has speeded up and moved forward relative to the light beam in the time that it takes the light to cross the room. Inside the laboratory, it will seem as if the beam of light is bent. It looks as if there is, after all, a way to distinguish acceleration from gravity. But no, says Einstein. We must keep the principle of equivalence unless it is proved false. If the light beam is bent in an accelerated laboratory (an accelerated **frame of reference**) then it must also be bent by gravity, and by the exactly equivalent amount. Since light has no mass, how can it be affected by gravity? Einstein puzzled over this from 1911 to 1915 before coming up with a mathematical theory, the **general theory of relativity**, that explained light bending, and much more besides.

Ten years earlier Einstein had published three papers, at the age of 26, that would alone have ranked him among the half-dozen great pioneers of twentieth-century physics. Those papers dealt with the **special theory of relativity**, how light is quantized into packets of energy (photons) and the way tiny particles move through the air or a liquid (**Brownian motion**). But all three topics were in the mainstream of physics at the time. Significant though Einstein's work was, if he had not done it then before long other researchers would have reached the same conclusions.

That is not true of the general theory of relativity. It was not a response to any observational puzzle (although it did, almost as an afterthought, account for an old mystery about the orbit of Mercury). Einstein was motivated by a deeper philosophical need, the quest for simplicity and unity in nature. If it had not been for Einstein a comprehensive theory of gravity might not have been developed for decades, until scientists were pressed to consider the need for such a theory by the discovery of objects such as black holes, pulsars and quasars.

Relativity visualized
Bending space and time

The new picture of the Universe casts aside the everyday notion of empty space and replaces it by an almost tangible continuum in four dimensions (three of space and one of time) that can be bent and distorted by the presence of material objects. It is those bends and distortions that provide the 'force' of gravity, bend light beams, and deflect moving objects from straight-line trajectories, a situation summed up by the aphorism 'matter tells space how to curve, space tells matter how to move'.

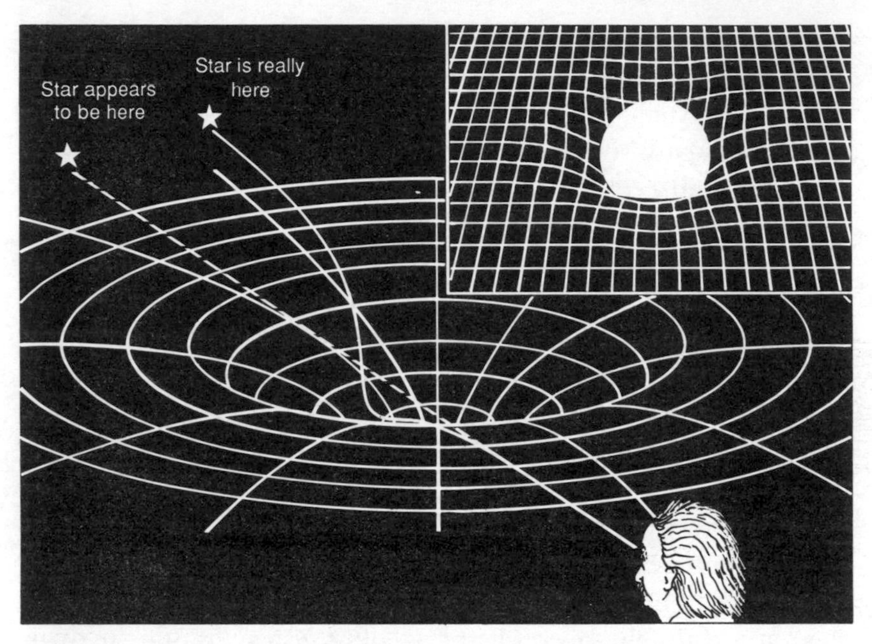

A heavy object placed on a stretched rubber sheet (insert) makes an indentation. The presence of the Sun 'indents' space-time in an analogous way. So light from a distant star is deflected as it passes the Sun

It is easier to visualize what is going on in terms of a two-dimensional elastic surface. Imagine a rubber sheet stretched tightly across a frame to make a flat surface. That is a 'model' of Einstein's version of empty space-time. Now imagine dumping a heavy bowling ball in the middle of the sheet. It bends. That is Einstein's model of the way space distorts near a large lump of matter. When you roll a marble across the original flat sheet it makes only a tiny indentation, and rolls in a straight line. But when you roll the marble near the bowling ball, the distortion in the rubber sheet makes it follow a curved path. That is Einstein's model for the force of gravity. Objects are simply following a path of least resistance, a **geodesic** – the equivalent of a straight line through a curved portion of space-time. And this explains light bending. The effect is the same for a marble, a planet, or a beam of light. When it moves near a large mass – through a gravitational field of force, on the old picture – it follows a curved trajectory.

General relativity predicted exactly how much a beam of light should get bent when it passes near the Sun, in order to be exactly equivalent to the bending seen by a physicist in an accelerated frame of reference. The new theory made a clear and testable prediction, that stars observed 'near' the Sun on the sky during a total eclipse (but actually, of course, much further away along the line of sight) would be displaced by a certain amount compared with their observed positions at other times. The prediction was confirmed by observations made during a total eclipse of the Sun in 1919.

General relativity has also passed every other test applied to it. One of these concerns what is known as the precession of the orbit of the innermost planet. Mercury, the closest planet to the Sun, orbits where the gravitational field is strong (that is to say, where space-time is strongly distorted). Astronomers already knew in 1915 that the orbit has a curious behaviour, which cannot be completely explained by Newton's theory of

During the solar eclipse of 1919 a team led by the English physicist Arthur Eddington measured the positions of several stars, shown by dots in the left diagram, which lay in nearly the same direction on the sky as the Sun (circle) at the time. Light from the distant stars passed through the region of space affected by the Sun's gravity. When Eddington compared these positions with the measured positions of the same stars when the Sun was on the opposite side of the sky, he found that they were apparently deflected. Each appeared to have moved by an amount which depended on the angular separation of the star from the Sun at the time of the eclipse. Light from each of those stars had been 'bent' as it passed by the gravitational field of the Sun. These 'deflections' (dots in right figure) fell exactly on the curve predicted by Einstein's theory

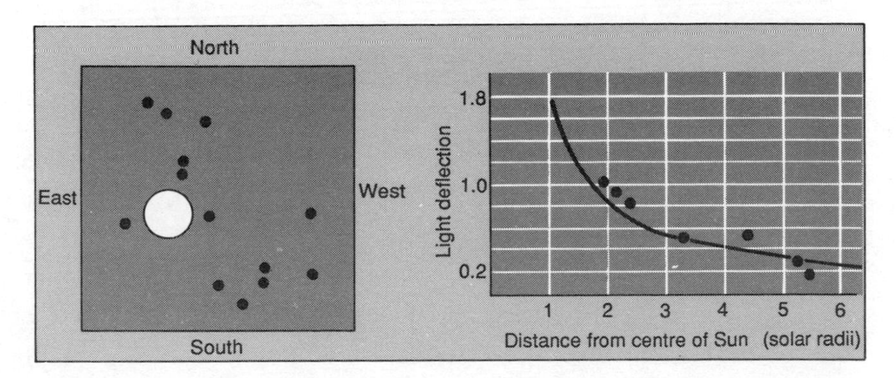

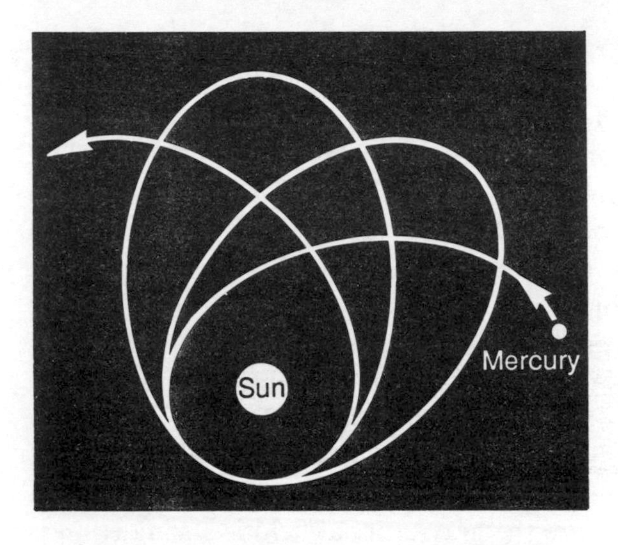

The curious shape of the orbit of Mercury (exaggerated in this illustration) is explained by general relativity

gravity. As Mercury follows an elliptical orbit around the Sun, the ellipse itself shifts slightly each orbit, tracing out a pattern like a child's drawing of the petals of a daisy.

This shift is exactly explained by general relativity. Anywhere that gravity is weak, general relativity and Newton's famous inverse square law – which says force is proportional to the inverse square of the distance between two masses (see Chapter 2) – give the same answers to the appropriate calculations. But in a strong field, according to general relativity, gravity deviates from the precise inverse square law. The size of the 'post-Newtonian' effect is just big enough at the distance of Mercury from the Sun to produce the puzzling changes in orbit.

The Universe at large
Einstein's great triumph

General relativity is a geometrical theory. It gives a well-defined physical meaning to a completely specified geometry of matter, space and time. But 'completely' is a key word here. In his search for a unified description of nature Einstein developed a theory that completely describes the Universe – and which, strictly speaking, *only* describes the complete Universe (or a complete universe).

When general relativity is applied to 'local' problems such as calculating the orbits of planets in the Solar System, it is being used in an approximation. In practice, such approximations can be made as accurate as you like, using **boundary conditions** to join the equations describing a local object like the Sun on to the rest of the Universe. But the point is that Einstein did not have to expand his theory to make it capable of dealing with the whole Universe – making it a cosmological theory. General relativity, from its birth, dealt quite happily with the whole Universe.

When Einstein tried to describe the simplest possible mathematical model of the Universe using his new equations, however, he ran into a problem. At that time, in 1917, the received wisdom was that our Milky Way Galaxy was the entire Universe, a stable collection of stars. But the equations describing a complete cosmology of space, time and matter refused to produce such a picture. They insisted that the Universe must either be expanding or contracting.

The only way Einstein could hold the model universe still, to mimic the appearance of the Milky Way, was to add an extra term to the equations, called the **cosmological constant**. In 1917 he wrote, 'that term is necessary only for the purpose of making possible a quasi-static distribution of matter, as required by the fact of the small velocities of the stars'. A dozen years later, observers led by the pioneer Edwin Hubble in California found that the Milky Way was not the entire Universe but simply one galaxy among many millions, and that distant galaxies are all receding from one another.

The Universe *is* expanding, exactly as the pure equations of general relativity predicted in 1917 when Einstein himself refused to believe the evidence of his own theory. There is no need for the cosmological constant, and Einstein's equations now provide the basis for the highly successful big bang description of the birth and evolution of the entire Universe (see Chapter 1).

Exotic phenomena
Relativity up to date

Within the expanding Universe, general relativity is required to explain the workings of exotic objects where space-time is highly distorted by the presence of matter – on the old picture, where large masses produce strong gravitational fields. The most extreme version of this, and one that has caught the popular imagination, is the phenomenon of **black holes**.

The concept of black holes is so familiar today, as a feature of Einstein's masterwork, that it may come as a surprise to learn that although the name black hole was first used in an astronomical context (by John Wheeler of Princeton University) only in 1968, the concept goes back more than two centuries, to the British polymath John Michell. Michell realized that because the speed of light is finite, and because the speed needed to escape from an object (the **escape velocity**) is greater for larger bodies, there must come a point where not even light can escape from the surface of a 'star'. This would be achieved by packing 100 million Suns alongside one another in a huge sphere.

Low-density black holes with 'only' the density of our Sun (1410 kilograms per cubic metre – a quarter of the Earth's density), or even less, may indeed exist in our Universe. If so, they would trap light by their gravitational pull – or, in terms of general relativity, by bending space-time around themselves so much that it becomes closed, pinched off from the rest of the Universe. But there is another way to make a black hole, which was first recognized as a mathematical possibility in the 1930s. If a star keeps the same mass but shrinks inward, or stays the same size while accumulating mass, density increases. Eventually, the distortion of space-time around it increases until, once again, a situation is reached where the object collapses and folds space-time around itself, disappearing from all outside view. Not even light can escape from its gravitational grip, and it has become a black hole. The notion of such stellar-mass black holes seemed no more than a mathematical trick, something that surely could not be allowed to exist in the real Universe, until 1968, and the discovery of **pulsars**.

Scientists now know that pulsars are the rapidly spinning remains of dead stars. They contain about as much matter as our Sun packed into a volume no bigger than that of a large mountain on Earth. Such **neutron stars** have roughly the density of the nucleus of an atom, and at 2×10^{17} kilograms per cubic metre are very close to the critical density at which gravity would overwhelm them and they would collapse into black holes. A neutron star could gain enough extra mass to do the trick by accumulating matter from interstellar space, or by stripping gas from a companion star by tidal forces.

The discovery of neutron stars made the possibility of black holes respectable. In the 1970s, several objects were discovered that might mark the locations of black holes. An object which emits no light (or anything else) cannot be observed directly. But a black hole orbiting around another star, and swallowing gas that it is tearing off its companion, would be a messy eater. The gas funnelling down into the black hole will get hot as the particles in the gas are accelerated and bash against one another. Astrophysicists calculated that the particles would get hot enough to radiate at X-ray frequencies. Scientists have now discovered X-ray sources in binary systems with the right properties to match those predicted for black holes by the equations of general relativity.

With black holes made respectable by these discoveries, they were soon invoked to explain another puzzling discovery of the 1960s, the **quasars**. Quasars are the energetic cores of some galaxies, which produce enormous amounts of energy from a region of space no bigger across than our Solar System. Allowing matter to fall into a strong gravitational field – converting gravitational potential energy into heat – is the most efficient way to produce energy, apart from the annihilation of particles with their antiparticle counterparts. Dropping a mass, *m*, into a black hole from infinite distance would release almost half of its **rest mass energy**, mc^2. If only a few per cent of this available energy is actually released when mass falls into a black hole, the energy needed to power a quasar could be provided by a big black hole which swallows just one or two times the mass of our Sun each year. The kind of black hole involved would contain about 100 million times the mass of our Sun – very much the sort of object envisaged by Michell two centuries ago. This would be no more than 0.1 per cent of the mass of all the stars in the galaxy surrounding the quasar. Such a black hole could arise simply because too many stars got too close together in the core of a galaxy.

A large concentration of mass will also bend light (that is, bend *space-time* so that light follows a curved path) near it, even without being a black hole. In some cases the mass concentration can act as a lens, focusing light from a distant galaxy or quasar to produce two (or more) images on the sky. Astronomers have now found such **gravitational lensing** in the Universe, where multiple images of a single quasar occur as a result of lensing by an intervening cluster of galaxies.

But the most impressive and complete proof of the accuracy of Einstein's theory comes from yet another phenomenon, **gravitational radiation**. The existence of gravitational radiation depends on Einstein's concept of space-time as a real, physical phenomenon which can be distorted by the presence of matter. The distortions are similar conceptually to the way in which a lump of matter when dropped into a pool of water makes waves on the surface of the water.

Ultimate proof
Gravity rules the waves

The image of matter as solid lumps embedded in a stretched rubber sheet, space-time, makes the origin of gravitational waves clear. When one of the lumps vibrates, it sends out ripples through the sheet, and these ripples set other lumps of matter vibrating. This is like the way a vibrating charged

Prospects for catching gravitational waves

A system such as the binary pulsar is like an extreme version of a rotating weightlifter's barbell. Viewed in the plane of rotation, this produces gravitational waves, which can be visualized in terms of their effect on a circular ring. Physicists call this kind of radiation 'quadrupole radiation'.

Quadrupole radiation can be understood most simply in terms of radiation from electric charges. A pair of electric charges, one positive and one negative, forms a dipole, and when these two charges move (vibrating or rotating) they produce dipole electromagnetic radiation. A dipole itself is electrically neutral overall. A pair of dipoles is a quadrupole (with two positive and two negative charges), and when the charges in such an array move (for example, with one dipole

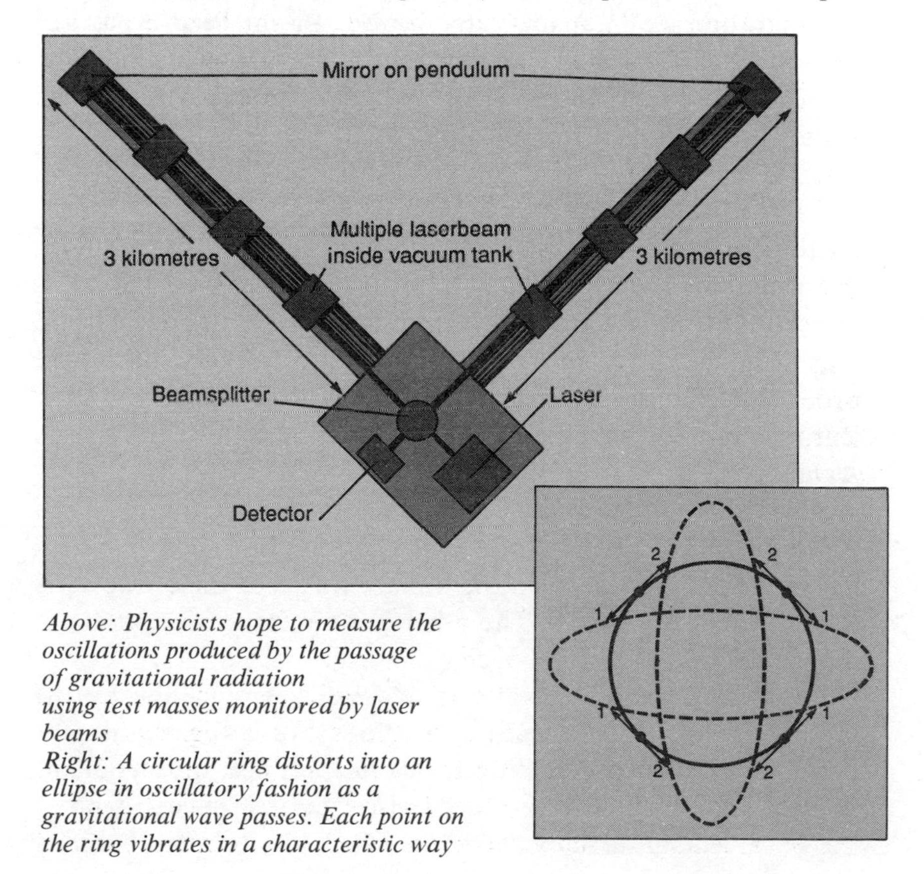

Above: Physicists hope to measure the oscillations produced by the passage of gravitational radiation using test masses monitored by laser beams
Right: A circular ring distorts into an ellipse in oscillatory fashion as a gravitational wave passes. Each point on the ring vibrates in a characteristic way

orbiting around the other) they produce quadrupole radiation.

Unlike electricity, mass comes with only one 'sign', so there is no gravitational equivalent of electric dipole radiation. Two masses which rotate around each other actually behave like a pair of dipoles, producing gravitational quadrupole radiation, which can be visualized in terms of its effect on a circular ring.

As the wave passes through, the ring is stretched in one direction and squeezed in another at right angles, becoming an ellipse. Then the pattern reverses. The pattern of alternate squeezing and stretching in two directions at right angles is the characteristic signature of quadrupole gravitational radiation.

Three test masses, placed in a right-angle 'L' shape, could detect such radiation as it squeezes and stretches space-time.

Such systems are now being constructed, with heavy masses placed in evacuated tubes several kilometres long, using laser interferometers to measure their positions to an accuracy of 10^{-18} metres. Researchers expect to detect gravitational radiation with such 'telescopes' during the 1990s.

---

particle sends out electromagnetic waves which shake other charged particles. But gravitational radiation is very weak – only 10^{-40} times as strong as electromagnetic radiation.

Researchers hope to measure the tiny ripples in space-time produced by massive objects far from Earth in the near future. But they already have proof that gravitational radiation exists. A binary system with two very dense stars orbiting rapidly around each other would, according to the equations, be a powerful source of gravitational radiation. Just such a system has been found. It is called the 'binary pulsar'. One of the stars in the system is a pulsar. The other is a neutron star that is not a radio source. They orbit each other every 7.75 hours.

Pulsars are superbly accurate 'clocks', keeping time by the sweep of their radio beams, like those of a lighthouse, as the neutron star rotates. Variations in the pulse rate from the binary pulsar show how the pulsar moves in its orbit – the observed pulse rate speeds up when the pulsar is moving towards us, and slows down when it is moving away. This is a version of the Doppler effect.

The period of the binary's orbit is slowly decreasing. This means that the two neutron stars are getting closer together as time passes. The reason is that the system is losing energy, in the form of gravitational radiation. General relativity predicts that the period of the binary pulsar should decrease by 75 millionths of a second each year; observations are so precise that they show a decrease of 76 ± 2 millionths of a second a year.

This is one of the greatest triumphs of Einstein's general theory of relativity. That theory is now established beyond any doubt as the best one that we have to explain gravity and the Universe at large.

24 February 1990

Further Reading

The best guide to Einstein's masterwork for the non-specialist is *Was Einstein Right?* by Clifford Will (Basic Books, 1986). The cosmological implications of general relativity are described in *In Search of the Big Bang* by John Gribbin (Corgi, 1987); Kate Charlesworth and John Gribbin tell the story of the Universe graphically in *The Cartoon History of Time* (Cardinal, 1990). Einstein's overall contribution to science is assessed in *Subtle is the Lord* by A. Pais (Oxford University Press, 1982), and in *Einstein: a Centenary Volume* edited by A. P. French (Harvard University Press, 1980).

CHAPTER 6

Life of a Star

Nigel Henbest

When we look up at the stars in the sky, we get the impression that they are changeless. Certainly the sky we see today is not very different from the view that our ancestors had 5000 years ago when they first connected up the stars into constellations: Ursa Major, the great bear; Taurus, the bull; Cancer, the crab; and so on.

But the stars do change. Rather like human beings, they are born; they live; and they die. A star's lifetime is very long compared with ours – many millions of years – so we only rarely see individual stars changing before our eyes. Astronomers can, however, work out the story of a star's life by picking out stars at different stages of their lives. An analogy can give an idea of the method. Imagine a Martian who lands on Earth in a crowded shopping centre. It stays for only a few minutes, so that nobody ages in front of its eyes. But it sees human beings at many stages of development:

The way a star lives and dies depends mainly on its mass: heavyweight stars have shorter lives and more spectacular deaths

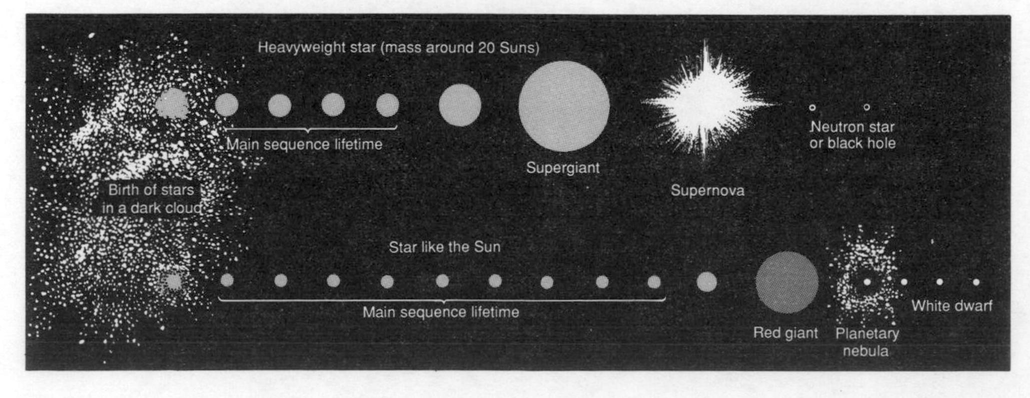

babies, old people, children, the middle aged, adolescents, pregnant women. From these data the Martian could work out how humans are born, live and eventually come to the end of their lives.

Similarly, astronomers can pick out stars that are being born, stars in the prime of life and stars that are dying. How do they know what stage a star has reached in its life? The answer comes from a detailed theory of star life, which is based on the laws of physics. This theory of **stellar evolution** is one of the great achievements of science in the twentieth century.

From the infinity of space
A star is born

Stars are born from the tenuous gas which fills the whole of space. This gas is composed mainly of hydrogen atoms, with a sprinkling of helium. In some places the gas clumps together in rather more dense **interstellar gas clouds**. According to gravitational theory the gas cloud's own gravity makes it attract itself. This should pull it in on itself, compressing the cloud to ever higher densities. The centre of the cloud should be the most compressed region. Here, astronomers expect some of the gas to condense into individual 'blobs', each held together by its own gravity. When a gas is compressed, it becomes hotter. So, the temperature at the centre of each blob rises to 10 million °C – hot enough to start nuclear reactions. These reactions turn hydrogen to helium and create vast amounts of energy. As a result the blob begins to shine: a star is born.

Unfortunately, ordinary telescopes cannot actually show us stars being born in the interstellar gas clouds. The problem is **dust**. Mixed in with the gas in space are small particles of dust – similar in size to the particles in cigarette smoke. In the denser clouds of gas, where the dust is more concentrated, dust particles absorb light passing through the cloud. As a result, we can see the clouds as dark silhouettes against a background of distant stars. The most famous dark cloud is the Coalsack,

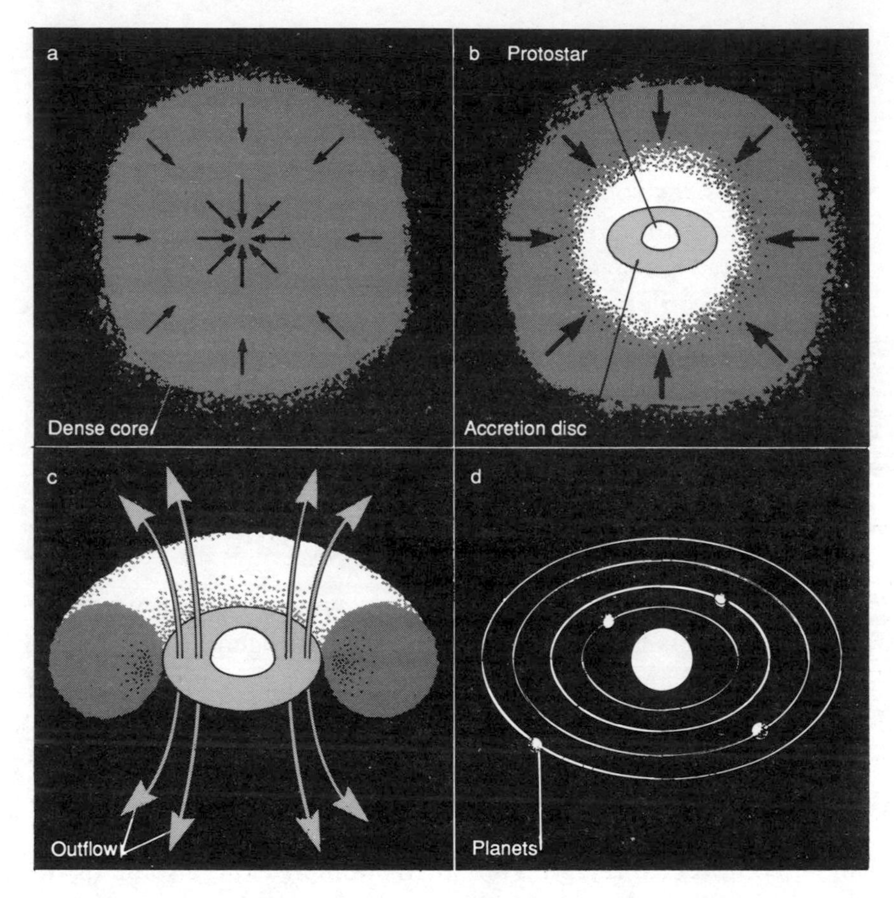

Birth of a star. A collapsing cloud of gas and dust (a) forms a dense core (a protostar) surrounded by a disc of gas and dust (b). Outflows of hot gas then drive away the remains of the original cloud (c). As nuclear reactions begin in the protostar, it becomes a star, and the matter in the disc condenses into planets (d)

which is visible to the naked eye from the southern hemisphere. The dust also prevents us from seeing what is happening inside the dark cloud – the region where stars are being born.

In recent years astronomers have solved this problem. They have built telescopes that pick up infrared radiation instead of light. The dust particles in space do not absorb infrared radiation, so the infrared telescopes can pick up radiation coming from within the dense clouds, and can 'see' the stars being born there. The most successful infrared telescope was on board a satellite which was launched into orbit in 1983. The Infrared Astronomical Satellite (IRAS) found thousands of young stars hidden deep within the interstellar clouds.

Astronomers have found that the blobs of gas collapse in rather an odd way. The central parts of a blob fall inwards rather quickly, while the outer parts follow at a more leisurely rate. The blobs are also rotating, quite slowly, but as the outer parts fall inwards, they begin to spin more rapidly – just as ice-skaters spin more quickly when they draw in their arms. As a result, the infalling gas forms a disc around the newly born star at the centre, where the gas is compressed enough for nuclear reactions to start. Within this disc, the gas and the dust that is mixed in it eventually form into a set of planets orbiting the new star.

Once the star is shining, it produces a powerful 'wind' of hot gas that forces its way outwards in opposite directions, above and below the disc. This wind drives away most of the original gas cloud that hides the star from view. Now we can see the young stars with an ordinary telescope. They light up the final tatters of gas from the original cloud, making it glow as a bright **nebula**.

Nebulae, each surrounding a 'nursery' of young stars, form some of the most beautiful sights in the sky. Most famous is the Orion Nebula, which you can spot from Europe with the naked eye during the winter months as a misty patch below the stars of the belt in the constellation of the mighty hunter.

In the prime of life
The main sequence

When a star is born it is a ball of hot gases, composed mainly of hydrogen. It shines because nuclear reactions at its centre are turning hydrogen into helium. To this extent, all new-born stars are the same. The main thing that marks out one star from another is its mass – the amount of matter it contains.

The mass of a star is fixed at its birth and it determines both a star's lifetime and its ultimate fate.

Our Sun is a very typical star, currently in the prime of its life, and so it makes a convenient yardstick for measuring other stars. Instead of saying a star weighs 20 000 million million million million tonnes, for example, we can say it is as massive as 10 Suns. On this scale, new-born stars cover a wide range, from as light as 0.07 Suns to as heavy as 100 Suns. Nuclear reactions run fastest in the heaviest stars, because their centres are the hottest and most compressed. So the heavier stars are the brighter stars, with hotter surfaces. We can arrange these stars in a definite sequence, called the **main sequence** of star types. At one end are the lightweight stars, which are much dimmer than the Sun and with a surface temperature of only 3000 °C. The Sun is in the middle, with a temperature of 6000 °C. At the top end of the range are heavyweight stars shining as brightly as 100 000 Suns, and with a surface temperature of 30 000 °C or more.

A star spends most of its life turning hydrogen into helium, so the duration of the main sequence period is really the prime of its life. The length of its life depends very critically on how heavy the star is. A heavyweight star uses up its nuclear fuel so rapidly that it soon exhausts its supplies of hydrogen. A lightweight star, even though it has a smaller supply of fuel to start with, uses it much more gradually, and so lasts for a much longer time. A star's lifetime is too long for us to appreciate easily, so again we can use the Sun as a comparison. According to theory, the Sun will spend 10 000 million years altogether as a main sequence star. The heaviest stars survive for only one thousandth of this time. The very lightweight stars can last for 100 times longer than the Sun.

When a star like the Sun dies, it doesn't just fizzle out. Instead, it experiences a kind of 'middle-age spread', and expands to become a **red giant**, about 100 times its previous size. The reason for this behaviour lies in the star's very centre, its core. Reactions here have turned the original hydrogen into helium. Like the ashes in a fire, this central region produces no

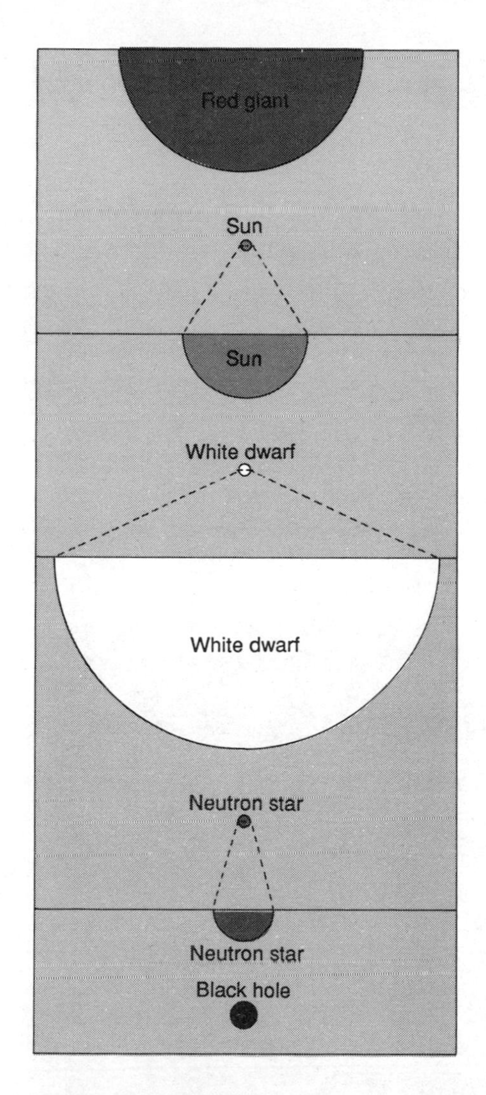

Stars range in size from black holes a few kilometres across to red giants a billion kilometres in diameter

energy. Nuclear reactions are still going on in a thin 'shell' around the helium core, and calculations show that these reactions produce more energy than before. As this extra energy pushes up through the star, it makes the outer parts of the star swell up. As its outer layers cool down, the star shines a red colour, hence the name 'red giant'. If we could cut a section through a red giant, we would find that it has a very small and dense core and a huge outer region of very thin gas – much more rarefied than the Earth's atmosphere.

Classifying the stars

The stars in our skies have a bewildering range of properties: there are giant stars and dwarf stars; bright stars and dim stars; hot stars and cool stars. Astronomers make sense of the stars by plotting them on a graph. A Danish astronomer, Ejnar Hertzsprung, and an American, Henry Norris Russell, found the most useful kind of graph back in 1914, and astronomers still call it

Astronomers plot the positions of stars on a 'Hertzsprung–Russell diagram' according to their luminosity (vertical scale) and temperature (horizontal scale). Most stars fall on the 'main sequence'; there are a few red giants (upper right) and white dwarfs (lower left). Our Sun can be seen to be a main sequence star. As a star ages, its plot would, in effect, move upwards and to the right on the diagram

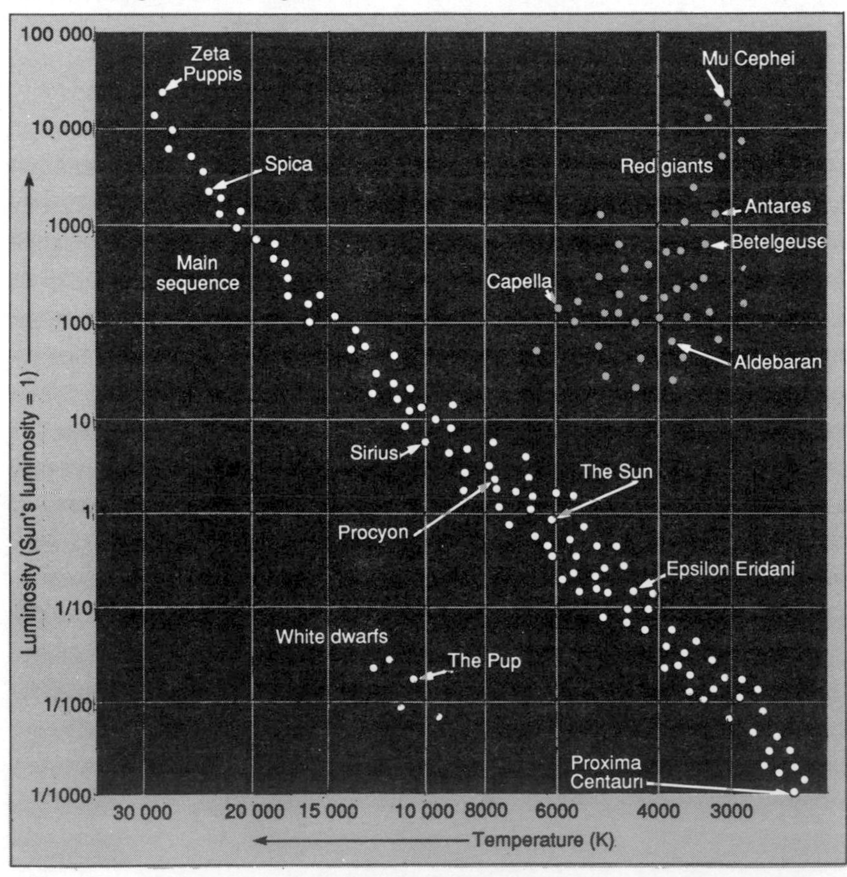

a Hertzsprung–Russell, or H–R, diagram.

On an H–R diagram the vertical axis represents the luminosity or brilliance of a star. The horizontal axis represents the star's temperature: for historical reasons this is plotted with high temperatures to the left and low temperatures to the right.

When Hertzsprung and Russell plotted the positions of stars on their diagram, they found that most of them occupy a narrow strip on the graph running from top left to lower right, the 'main sequence'. We now know that these are stars that derive their energy by turning hydrogen into helium.

Compared with the main sequence stars, red giant stars are not very common; however, because they are very large they appear bright and stand out conspicuously in our skies. The most famous is Betelgeuse in the constellation Orion; another is Antares, in Scorpio: the Greek name means 'the rival of Mars' because of its brilliant red colour. A red giant finds it difficult to hold on to its huge outer regions. The star becomes unstable and eventually the outer gas drifts off into space. Before completely disappearing the gas forms a bubble around the dying star – the effect is like a glowing smoke ring in space. Astronomers call these bubbles **planetary nebulae**, because they look rather like a planet when you observe them with a small telescope. After the star's outer regions have disappeared, we can see the tiny, very hot core. It is only one hundredth the diameter of the Sun – no larger than planet Earth – and is so hot that it shines white-hot. Astronomers call this a **white dwarf**.

Because white dwarfs are very small they appear as rather dim objects in the sky and thus are difficult to find. But astronomers have been highly successful in tracking down white dwarfs when they are a companion to another star. The first to be discovered was the companion to Sirius, the brightest star in the sky. Because Sirius is known as the Dog Star, its small companion is often called the Pup.

A white dwarf is no longer producing any energy. It shines merely because it began life so hot. As time passes, it gradually cools down, fading through yellow, orange and red, until – like a dying ember in a fire – it fades from sight altogether.

Supernova! An explosion in space

A heavyweight star has a much more dramatic end – as astronomers in the southern hemisphere saw in 1987. A star previously visible only through a powerful telescope suddenly exploded, and shone so brightly that it was easily visible to the naked eye. The star had died as a **supernova**.

The build-up to a supernova starts after a heavy star has lived out its main sequence of life. Now that the star has used up its central supplies of hydrogen, it expands to become a red giant, with a central compact core of helium. But this is not the end of the story. In the middle of such a massive star, the pressure and temperature keep on rising until helium atoms begin to fuse into a heavier element, carbon. This reaction produces extra energy to keep the star shining. Eventually, the increasing temperature and pressure force the carbon to change to even heavier elements, such as neon, silicon and iron.

At this point the star's core is like an onion, with concentric layers (from the inside out) of iron, silicon, neon, carbon, helium and hydrogen. But the process cannot carry on indefinitely. If you try to fuse together iron nuclei, the reaction does not produce energy – in fact, it takes in energy. So the star's centre is now unstable. In just a few seconds, it collapses entirely. A wave of energy from the collapsing core blows the star apart, in the explosion of a supernova.

Neutron stars and black holes

But what happens to the collapsing core of a supernova? In the 1930s two astronomers working in the US, Fritz Zwicky and

Walter Baade, suggested that it shrank into a small ball, smaller than a white dwarf, made entirely of the subatomic particles called neutrons.

For decades this was just a theoretical idea – until one autumn day in 1967. Two radio astronomers at Cambridge, Tony Hewish and Jocelyn Bell, picked up regular signals coming from the sky. They dismissed the idea that it might be 'little green men' trying to contact the Earth, and realized that instead they had found some kind of natural lighthouse in space. The lantern of a lighthouse sends out beams of light that seem to flash as the lantern rotates. The signals picked up at Cambridge must have come from a cosmic lighthouse that was

A supergiant star (below) has a small dense core composed of layers of different elements, like the shells of an onion. A black hole (bottom) can pull gas from a companion star. The gas emits X-rays as it spirals into the hole

A neutron star (below right) has a strong magnetic field which generates beams of radiation. As the beams sweep past the Earth, radio astronomers pick up regular pulses of radio waves

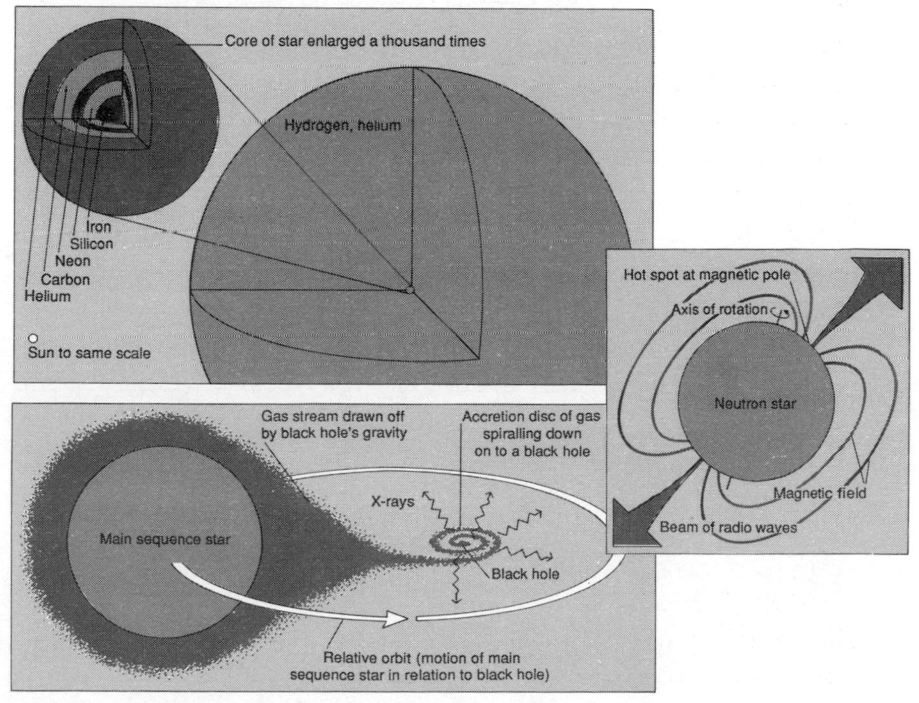

emitting beams of radio waves, and spinning about once a second. From our present knowledge only one kind of star was small enough to spin so rapidly – a neutron star.

Radio astronomers have now located hundreds of spinning neutron stars (also known as **pulsars**, because of their regular 'pulses' of radio waves). One of them lies at the centre of the Crab Nebula, the twisted gas cloud thrown out by a supernova that exploded 900 years ago. A neutron star is only about 25 kilometres across, and the material inside it is so tightly packed that a pinhead of matter from a neutron star would have a mass of a million tonnes. Its gravity is so strong that an astronaut who tried to land on its surface would be crushed and spread out to form a layer only one atom thick!

White dwarfs and neutron stars may seem very bizarre, but theory predicts an even odder type of 'star corpse': a **black hole**. If the collapsing core of a supernova is too massive (heavier than three Suns), it cannot end up as a neutron star. Its own gravity is so powerful that the core continues to shrink, until it becomes a mathematical point, with no size at all and an infinite density. Surrounding this point is a region a few kilometres across where gravity is so strong that nothing can escape – not even light. This region is a black hole. It is 'black' because it does not let light escape; and even if you tried to illuminate it, the hole would swallow up the beam from your torch. It is a 'hole', because anything you throw into it can never emerge again, however powerful the rocket engines you might strap to it.

As with neutron stars, astronomers first predicted black holes in the 1930s. Only in the past few years have they found some evidence for them. In the constellation Cygnus (the swan), there is a powerful source of X-rays, named Cyg X-1. Astronomers have found a star at this point in the sky. The star itself is quite ordinary, and cannot be producing the X-rays. But it is not on its own. It is swinging around a companion star that is invisible in ordinary telescopes. By observing the visible star carefully, astronomers found that its invisible companion was exerting the gravitational pull of an object as heavy as 10 Suns. This is much too heavy to be a neutron star, and so the only possibility is that it is a black hole.

Rebirth! Cocktail of new elements

Supernovae do not only represent death and destruction. The blast from a supernova sweeps up the gases in space, compressing them into dense clouds. Here gravity can get to work, making the gas clouds shrink and condense into a new star. So a star is like a phoenix: the death of a star as a supernova can trigger off the birth of a new generation of stars.

When a star dies – as a planetary nebula or a supernova – it seeds space with the new elements that it has created during its lifetime or in its death throes – elements such as carbon, iron, gold and even uranium and other radioactive elements. So the newly born stars will contain slightly less hydrogen, and rather

Life and death of the Sun

Our knowledge of other stars gives us a means of predicting the fate of the star that is most important to us: the Sun. A comparison with other stars with the same mass tells us that the Sun is a main sequence star, shining because it is turning hydrogen into helium at its centre.

The Sun was born about 5000 million years ago, and we can predict that it has enough hydrogen to carry on much as it is for another 5000 million years. Then it will begin to swell.

As the Sun turns into a red giant, it will swallow up Mercury, and then Venus. The bloated Sun will boil away the Earth's oceans, destroying any life that has not fled to another planetary system. Then the Sun will engulf the Earth itself.

Eventually, the Sun's outer layers will puff away as a beautiful planetary nebula, leaving a white dwarf at the centre of what is left of the Solar System, circled forlornly by the charred remnants of its remaining planets.

more of these exotic elements. Astronomers now believe that when the Universe began, in the big bang, the gases consisted almost entirely of hydrogen and helium. Dying stars have formed all the other elements, including the silicon, oxygen and iron that form the Earth, and the carbon and other elements in our bodies. So we owe our very existence to the life and death of countless past generations of stars.

2 June 1988

Further Reading

100 Billion Suns by Rudolf Kippenhahn (Weidenfeld & Nicolson, 1983) is an excellent introduction to the birth, life and death of stars. *Superstars* by David Clark (Dent, 1979) provides a popular level introduction to supernovae, written before the bright supernova of 1987. 'First light on starbirth' and 'Supernova: the cosmic bonfire' are articles in *New Scientist* that present the latest results on the birth and death of stars (27 August 1987, p. 46; 5 November 1987, p. 52).

CHAPTER 7

Origin of the Chemical Elements

Tony Cox

Everything that we see around us is made up of about 90 chemical elements. Their discovery and identification was one of the great achievements of chemistry in the eighteenth and nineteenth centuries. In the early part of this century there came a better understanding of the atoms characteristic of each element. Negatively charged electrons appeared to 'orbit' a positive nucleus, rather as the planets go round the Sun. The heavy atomic nucleus makes up nearly all the mass of an atom although, relative to the size of the whole atom, it is very tiny. This nucleus is composed of positively charged protons together with some neutrons, which have no electric charge. The chemical nature of an element is controlled by the number of protons, ranging from one in hydrogen to 92 in uranium, and even more in elements that are not to be found naturally on Earth, but which chemists have succeeded in making in recent decades.

Where do the chemical elements come from? Chemical reactions rearrange atoms into different combinations, making and breaking chemical bonds. Only the outer electrons of an atom will take part in this process: the central nucleus is unaffected. Such reactions are not able, therefore, to turn one element into another element. That requires rearranging the particles making up the nucleus. This happens in radioactivity, when unstable nuclei split up, sometimes making lighter ones (see Chapter 3). The nuclei of light atoms can also fuse together, making heavier atoms. This is a difficult process because the two nuclei coming together need a large energy to get close enough to 'stick'.

Physicists use high-energy particle accelerators to study this type of nuclear reaction.

But atoms can also fuse at very high temperatures, when they move around quickly and randomly. Chemical reactions require some energy of this kind, and generally speed up at higher temperatures. But those reactions happen at temperatures of tens, hundreds, or at most a few thousand, degrees. For nuclei to react, very much hotter conditions are needed – a minimum of 10 million °C.

A tunnel through to heavier nuclei

The protons and neutrons inside nuclei are 'glued' together by an attraction called the **strong force** (see Chapter 3). An important feature of this force is that it operates over exceedingly short distances, around 10^{-13} centimetres, about the size of nuclei themselves. There is another important force at play: the electrostatic repulsion between positively charged protons. Inside the nucleus, the strong interaction is sufficient to overcome this, and so strong enough to ensure the stability of nuclei.

Suppose we try to force two nuclei together, to make a heavier element. This process happens quite easily, if we can get them close enough together for the strong interaction to operate. But the electrostatic repulsion acts over much longer distances. So before the nuclei can get very close, a very large repulsion operates.

This gives rise to an energy barrier, known as the **Coulomb barrier**, which is shown in the diagram on the next page. In classical physics, two nuclei would have to have enough energy to surmount this barrier before the fusion reaction could take place. Quantum physics, however, introduces an important subtlety here. According to this theory, microscopic particles can pass through energy barriers which in classical physics are impenetrable. This process, known as **tunnelling**, is very important in many processes of radioactive decay; it is also essential in the fusion reactions that make heavy elements.

The amount of tunnelling depends on the energy of the particles, and at ordinary temperatures it is negligible. High

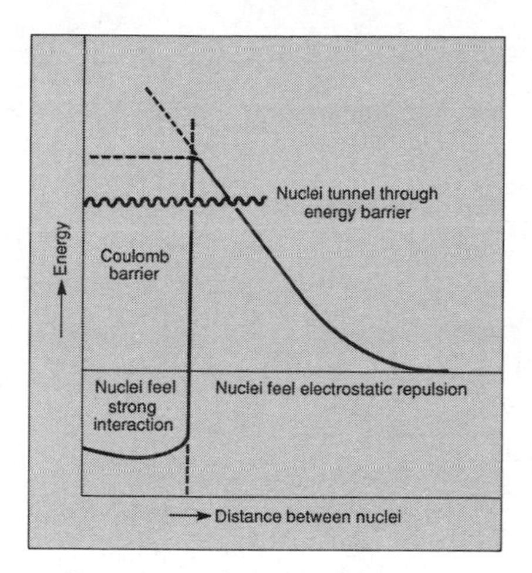

As the nuclei come closer together, they feel a repulsion due to their positive charges – the so-called Coulomb energy barrier. But the nuclei can 'tunnel' through the barrier, getting close enough to feel the strong nuclear force. At very short distances they can fuse to form a heavier, stable nucleus

temperatures, where atoms have large random velocities, are still required for fusion to take place; but if it were not for tunnelling, these temperatures would have to be much greater still, and the production of new elements would be much harder.

The size of the Coulomb barrier increases with the charge on the approaching nuclei. To fuse heavier elements, therefore, requires higher temperatures. Neutrons, however, have no electric charge, and so they are able to approach nuclei without any repulsion. So it is, therefore, much easier to make heavier nuclei by adding neutrons than it is by the normal fusion process.

Temperatures such as these existed in the very early stages of the Universe, in the first few minutes after the big bang – the fireball that started off the Universe some 15 billion years ago. Nuclear reactions also occur inside stars, and in fact nuclear reactions are the only possible source of the enormous amount of energy that keeps stars hot for billions of years. So the big bang at the beginning of the Universe and the interiors of stars provide the two environments where most elements have been made.

Abundance of elements
Analysing the stars

The elements on the Earth vary enormously in their abundance. Oxygen, silicon and iron are common; many elements, such as gold, are millions of times as rare. Most of the elements are important to us in various ways. For example, iron forms the major part of the Earth's core. The surrounding layers of our planet, including the crust, are made of silicon and oxygen together with many other elements in smaller proportions. Nearly 30 elements are essential to life. These include carbon, oxygen, hydrogen and nitrogen, but also some quite rare ones such as selenium (see Chapter 16). In this century industry has come to make use of nearly all the elements. The manufacture of a modern touch-dialling telephone, for example, will involve no less than 42 of them.

Many features of the Earth's chemical composition, however, are not at all typical of the Universe as a whole. We know that the two lightest elements, hydrogen and helium, make up more than 99 per cent of the visible Universe, with the others being present in very small proportions. On Earth, hydrogen, and especially helium, are much rarer because they are gases except at very low temperatures, and they largely escaped into space when the Earth formed. The common elements on Earth condensed into solids – metallic iron, and silicon oxides – and so became concentrated in the dust particles that eventually collected together to make up our planet.

The abundance of elements in space is very important, not only because it influenced the ultimate composition of the Earth, but also because it can provide many clues to how the elements were originally formed. Clearly, a satisfactory theory about the origin of the elements should be able to account for the abundances that we observe.

How do we know these abundances? The nineteenth-century philosopher Auguste Comte thought that it would be impossible ever to know the composition of stars and other bodies in space. But even as he was writing, the evidence needed for this was becoming available. If you measure the **spectrum** of sunlight, by splitting the light into different colours, or wavelengths, with a prism of diffraction grating, you see many dark lines running across. Chemists see the same lines in spectra produced in the laboratory, when they put different

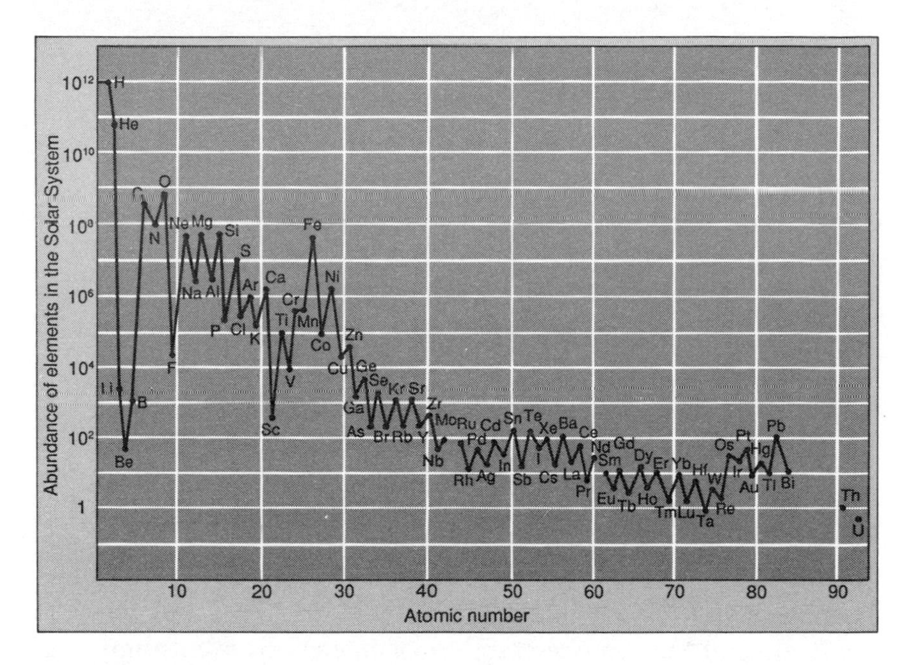

The relative abundances of elements in the Solar System. Elements are shown with their atomic number (number of protons) and chemical symbol. The graph shows the numbers of atoms present for every 10^{12} hydrogen atoms. The vertical scale is a logarithmic one, each scale mark representing a whole factor of 10 different from the next one. The 'zigzag' curve shows that elements with an even number of protons are more abundant than those with an odd number

elements in a flame. The lines are there because different elements absorb or emit light corresponding to characteristic, and extremely precise, wavelengths.

The wavelengths of the lines within the spectrum of the Sun reveal to us what elements are present in its outer layers, where atoms absorb some of the light coming from the interior. The strengths of these lines show us how much light is absorbed, and, therefore, the amount of each element that is present.

Astronomers have carried out this kind of analysis on the Sun and on many stars and galaxies, building up a picture of the abundance of elements in space. The information is supplemented by analysing meteorites, especially rare types called **carbonaceous chondrites**. These extraterrestrial rocks probably contain material left over from the formation of the Solar

System, which was not incorporated into the planets. The chemical composition of these meteorites matches that of the Sun very closely, except that they lack a few light elements such as hydrogen and helium, which did not condense into solids.

The combination of data from the solar spectrum and the chemical analysis of meteorites allows us to know fairly precisely what the overall composition of the Solar System is. The abundances found are shown in the diagram on page 83. Notice the scale and the enormous range of abundances: each scale mark in the abundance scale differs from the neighbouring one by a factor of 10. For every 10^{12} atoms of hydrogen there are about 10^{11} of helium, fewer than 10^9 of the next commonest elements, carbon and oxygen, and fewer than 10 each of some rare elements such as uranium. The 'zigzag' curve shows that elements with even numbers of protons tend to be more common than those that have an odd number, and this is a reflection of the relative stability of nuclei: an even number of protons or neutrons gives extra stability.

The first few minutes
Elements one and two

Analysing other stars shows that the Solar System is fairly typical in its composition. There is, however, an important observation: very old stars, which started life as long as 10 billion years ago (the Sun is less than half this age), are made of hydrogen and helium, with relatively much less of the heavier elements. This suggests that the heavier elements were even rarer when these old stars were formed. In fact, theorists suggest that only hydrogen and helium were made in the big bang. All the other elements must have been made since. Even today, they may be growing more abundant.

According to the big bang theory, the Universe began as a 'fireball' of extraordinarily dense and hot matter. In the early stages it was so hot that not even atomic nuclei – let alone the molecules and solids familiar in everyday life – could be stable.

The chemical elements could not have been present 'in the beginning', but must have been made subsequently. Some early speculations on the big bang theory suggested that all the known elements might have been produced very rapidly by putting light nuclei together in the early stages of the Universe. We now know that this was impossible, because the extremely rapid cooling and expansion of the Universe did not leave enough time. There was, however, some synthesis of elements in the first few minutes, and cosmological theory today can explain very nicely the apparent composition of the early Universe.

A few seconds from the beginning, the temperature was around 10^{10} °C. This is the maximum temperature at which atomic nuclei, other than simple protons, can exist. Protons were constantly changing into neutrons and vice versa, giving a ratio of about one neutron to every seven protons. Free neutrons are unstable, and under normal conditions last for only 15 minutes on average, then decay to hydrogen atoms (protons and electrons). Before this decay process, there was time for neutrons and protons to combine, forming deuterium, a heavy form of hydrogen. At the high temperatures that were then prevalent, the deuterium nuclei reacted rapidly with more protons and the ultimate product was the stable nucleus of helium, containing two protons and two neutrons. Under these conditions the proportion of helium formed relative to hydrogen depends on how many neutrons are available at the temperature where nuclear reactions can begin. Physicists can calculate this quite precisely, and the theoretical value – about one helium atom to ten of hydrogen, or 23–25 per cent helium by mass – agrees very well with the proportions found in the Universe, especially in the older stars. A small amount of deuterium was also left unreacted; the predicted abundance of it also agrees well with what scientists have observed.

Any further nuclear fusion reactions, making heavier elements from helium, could not happen to an appreciable degree because the temperature was too low by the time helium was made. It seems, therefore, that 99 per cent of material in the

Universe today owes its origin to the early stages. The agreement between theory and observation is impressive, and is one of the strongest pieces of evidence that ideas about the big bang are correct: no other theory of the origin of the Universe can explain the existence of hydrogen and helium in their observed proportions.

Cosmic cooking pots
The heavier elements

Although elements heavier than helium make up only 1 per cent of the Universe, they are essential to us in many ways. The very existence of solid planets, such as the Earth, depends on elements such as iron, silicon and oxygen. We are made of highly complex molecules that contain carbon, nitrogen and many other elements.

A universe made of hydrogen and helium would be a very dull place in chemical terms. It is impossible to imagine how intelligent beings could arise to observe it or write about it. To make the heavier elements requires high temperatures sustained over a much longer period of time than was so after the big bang. But such conditions do exist now – at the centre of stars. It is here that most of the remaining elements are made.

A star begins when a large mass of gas contracts under its own gravity. Compression raises the temperature in the centre, to the point at which nuclei can start to fuse to form heavier nuclei. The output of energy from the nuclear fusion keeps stars hot and prevents any further contraction, at least until the nuclear 'fuel' has been used up. The first reaction to begin, at a temperature of about 10 million °C, is the fusion of hydrogen nuclei (protons) to form helium; this reaction occurs in a number of steps, in some of which half the protons are converted into neutrons. This is the so-called **hydrogen burning**

Opposite: *Generations of stars exploding as supernovae produce the heavy elements (shown here as black dots) that are needed to form planets and people*

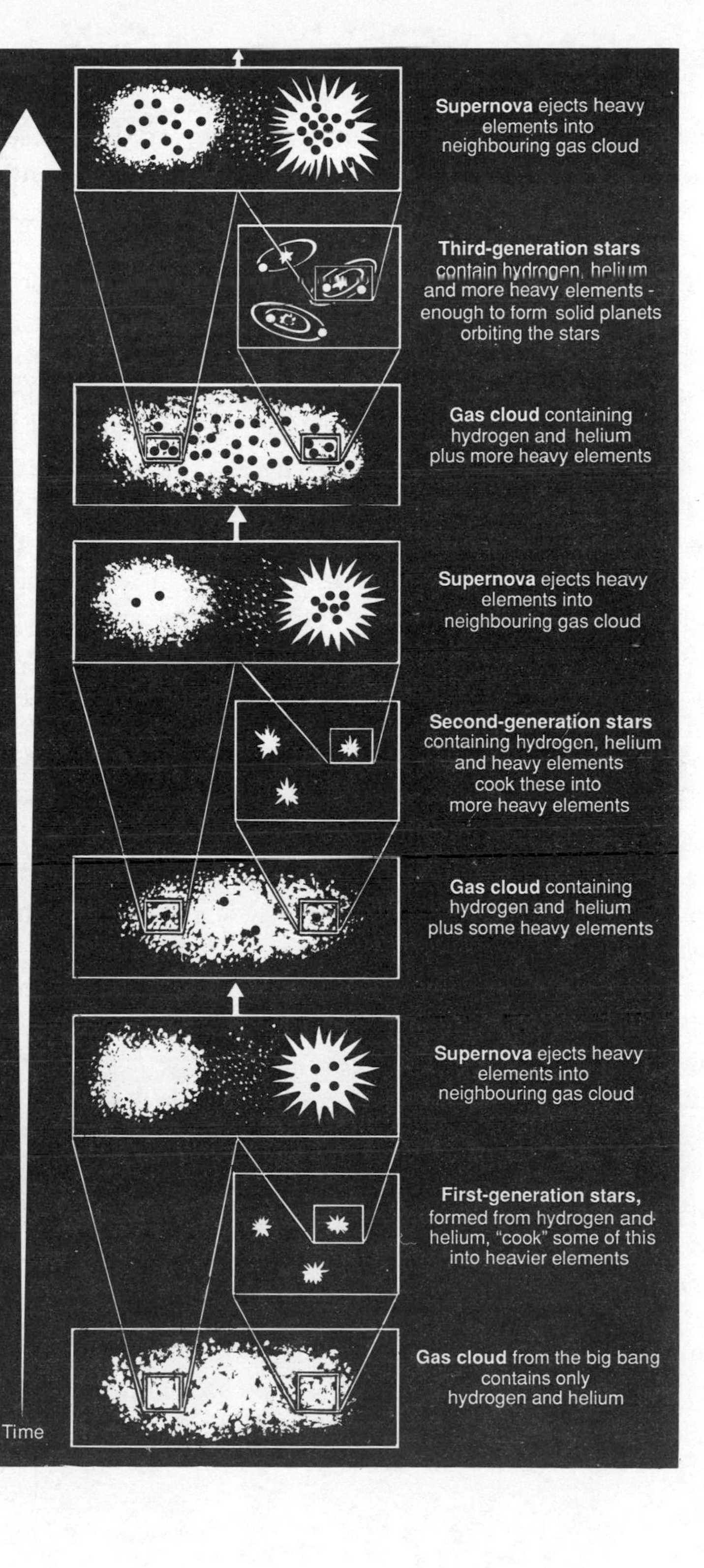
Supernova ejects heavy elements into neighbouring gas cloud
Third-generation stars contain hydrogen, helium and more heavy elements - enough to form solid planets orbiting the stars
Gas cloud containing hydrogen and helium plus more heavy elements
Supernova ejects heavy elements into neighbouring gas cloud
Second-generation stars containing hydrogen, helium and heavy elements cook these into more heavy elements
Gas cloud containing hydrogen and helium plus some heavy elements
Supernova ejects heavy elements into neighbouring gas cloud
First-generation stars, formed from hydrogen and helium, "cook" some of this into heavier elements
Gas cloud from the big bang contains only hydrogen and helium
Time

phase of stars. It is not burning in the everyday sense of the word. Hydrogen burning does not produce new elements, but it is important because the energy produced keeps stars going for much of their lives.

The hydrogen burning phase leads to the build-up of a core of helium in the centre of the star. When hydrogen is exhausted, and the output of energy from the reaction declines, the centre of the star starts to contract again and becomes even hotter. As the core shrinks, the outer parts of the star expand. The star grows into a red giant. What happens next depends on the star's mass. In the case of stars that have a relatively low mass, the core of helium simply becomes a compact object no larger than the Earth, known as a white dwarf, in which the helium nuclei are closely packed. The outer layers escape into space.

If a star is more massive than 0.4 Suns, the core becomes so hot (around 100 million °C) that the helium nuclei can react to form heavier nuclei. These fusion reactions require higher temperatures because the nuclei are more highly charged, and so need more energy to overcome their mutual electrostatic repulsion and fuse.

Two helium nuclei form beryllium (with four protons) but this nucleus is quite unstable and reacts quickly with further helium nuclei, to form first carbon and then oxygen. These two elements are the commonest in the Universe, after hydrogen and helium. The relative amounts made depend on the temperature of the star, which in turn is controlled by its mass. But astronomers also know that some subtle features of nuclear physics are involved. In fact, it is something of an 'accident' that carbon does not react so quickly as to be effectively bypassed by this sequence. A world without carbon would be one without us!

As helium is consumed, a core of carbon and oxygen builds up. For a star with a mass between 0.4 and 8 times that of the Sun, this is the end of fusion reactions. The core becomes a white dwarf that is composed of carbon and oxygen. In the most massive stars the core gets so very hot that carbon and oxygen can in turn fuse together, forming elements as heavy as

Jokers in the pack

Lithium, beryllium and boron (elements with, respectively, three, four and five protons) are comparatively rare. Their nuclei are not very stable and they are immediately consumed by nuclear reactions in stars. A little lithium probably came from the big bang, but most of these light elements are believed to have been made in a different way from the others through collisions with **cosmic rays**. These rays are mostly nuclei travelling through space at high speed. Their origin is still uncertain: some may come from supernovae, or from other high-energy events in the Universe. Their energy is so large, however, that when they collide with other atoms in space, the nuclei can break into very much smaller fragments.

This process, known as **spallation**, is probably the origin of most lithium, beryllium and boron. Evidence for this comes from the atomic composition of the cosmic rays themselves: they do, indeed, contain these elements in very much higher relative proportions than does the Solar System, or even the Universe as a whole.

---

sulphur. Further reactions happen in stages, eventually producing iron (which has 26 protons) and a number of elements with similar masses, right at the centre. The reactions stop here, because iron has the most stable nucleus of all elements and cannot fuse under these conditions.

Around the iron core there are various layers in the star where the other reactions are still going on, so in cross-section the star tends to resemble an onion. As well as the reactions that build hydrogen up to iron, other fusion processes are going on in these layers. These minor reactions can build up nuclei that are heavier than iron, in what astronomers call the ***s*-process** (meaning 'slow'). The *s*-process occurs when some reactions produce neutrons, which are captured by other nuclei, so increasing their weight. Once a neutron has been captured, it may change into a proton. In this way, the *s*-process can increase the number of both protons and neutrons within a nucleus. It can produce elements up to bismuth (which has 83 protons).

The death of a star
Elements of a supernova

The life of a star reaches its final stage when the core of iron builds up in the centre. The iron nuclei cannot produce energy by fusion, but the gravitational force is remorseless: it continues

The 'shell' structure of a heavy star, just before its explosion in a supernova, showing the main elements found in each shell. The diagram indicates the relative mass of each part, but not its size: the inner shells are much denser, and occupy very much less space than indicated

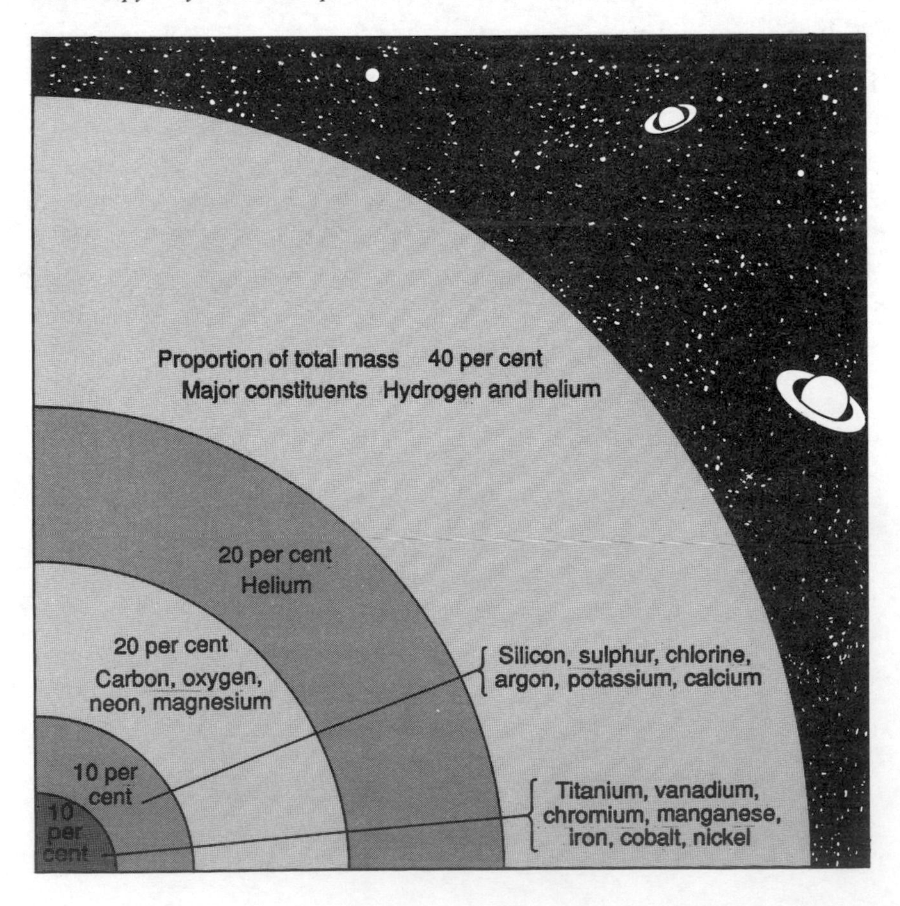

to compress the core, raising the temperature to billions of degrees. Some of the elements formed in the core begin to disintegrate in this inferno, and the very centre of the star collapses suddenly into a dense mass of solid neutrons. The outer layers fall in, then 'bounce back', spewing the contents of the star out into space in a **supernova** explosion. The explosion itself creates more heavy elements, because it produces a flood of neutrons that are absorbed by existing nuclei. Unlike the *s*-process, where neutrons add on to nuclei one by one, there are now so many neutrons that several attach to a nucleus at once. This ***r*-process** (for 'rapid') can make elements as heavy as uranium.

In a supernova explosion, the star becomes very much brighter, sometimes as brilliant as a billion Suns. Over the past 50 years, astronomers have found hundreds of supernovae in distant galaxies. These were so far away that they needed a telescope to be seen. When a supernova occurs in our Galaxy or a near neighbour galaxy, it is sometimes bright enough to be easily visible with the naked eye. We can find several supernovae in historical reports, including an observation by Chinese astronomers in AD 1054. As we saw in Chapter 6, the remains of this supernova now form the Crab Nebula, a cloud of hot gas still expanding outwards from the explosion. The spectrum of the expanding gas shows the presence of several elements made inside the star. The most recent supernova visible to the unaided eye was seen in 1987. The spectrum of its gases show many elements made in the explosion, including some that are radioactive and have gradually dwindled since 1987.

Some stars expel gas in more gentle ways, but supernovae provide the most important route for getting the elements out into space. Products from supernovae spread out, and eventually mix up with more gas. They then become incorporated into later generations of stars formed from the gas, eventually forming planets as well. Apart from direct observation of the remnants of old supernovae, the best evidence for the theory that the elements are produced in stars is that calculations confirm the observed abundances of elements. Such calculations are difficult and require the power of modern supercomputers. But the agreement is good. It appears from such calculations that almost all the material of the Solar System, apart from the hydrogen and helium remaining from the big bang, was produced by supernovae during the first few billion years of our Galaxy's existence.

3 February 1990

Further Reading

The Origin of the Chemical Elements by R. J. Taylor (Wykeham, 1975). *The Elements: their Origin, Abundance and Distribution* by Tony Cox (Oxford University Press, 1989) gives more details, and also discusses the elements on Earth. *The First Three Minutes* by Steven Weinberg (Fontana, 1983) remains the best account of the synthesis of elements in the big bang. *The Cambridge Encyclopedia of Astronomy* (Cambridge University Press, 1984) contains nicely illustrated articles on stars, including nuclear reactions and supernovae.

CHAPTER 8

Radioactivity

Christine Sutton

Many people fear radioactivity: they associate it with the fallout of atomic bombs or disasters such as the explosion at the Chernobyl nuclear power station. It is, however, a natural process, happening constantly all around us. It occurs within our homes, in the food we eat; even our bodies are radioactive. Much of this radioactivity is a natural consequence of the composition of the rocks and soil at the surface of the Earth. It is a faint remainder of the far greater radioactivity that existed when the Earth formed over 4000 million years ago. Indeed, without radioactivity, stars would not shine and the ingredients from which we and our planet are built would never have been formed.

Where our annual dose of radiation comes from

We are all exposed to **background radiation** from natural sources, no matter where we live, although the level varies from place to place. A person living in Britain typically receives an average annual dose of 2.18 millisieverts (mSv) from natural sources. This is 87 per cent of the total annual dose; the remaining 13 per cent comes from man-made sources.

The greatest contribution (1.3 mSv) to the natural background comes from isotopes of the radioactive gas radon, and the decay products. These isotopes of radon are a product of the transmutations of the small amounts of uranium and thorium in rocks, soil and building materials.

In some regions, such as South-west England, levels of radiation exposure from these isotopes are several times greater because the local granite contains more uranium than elsewhere.

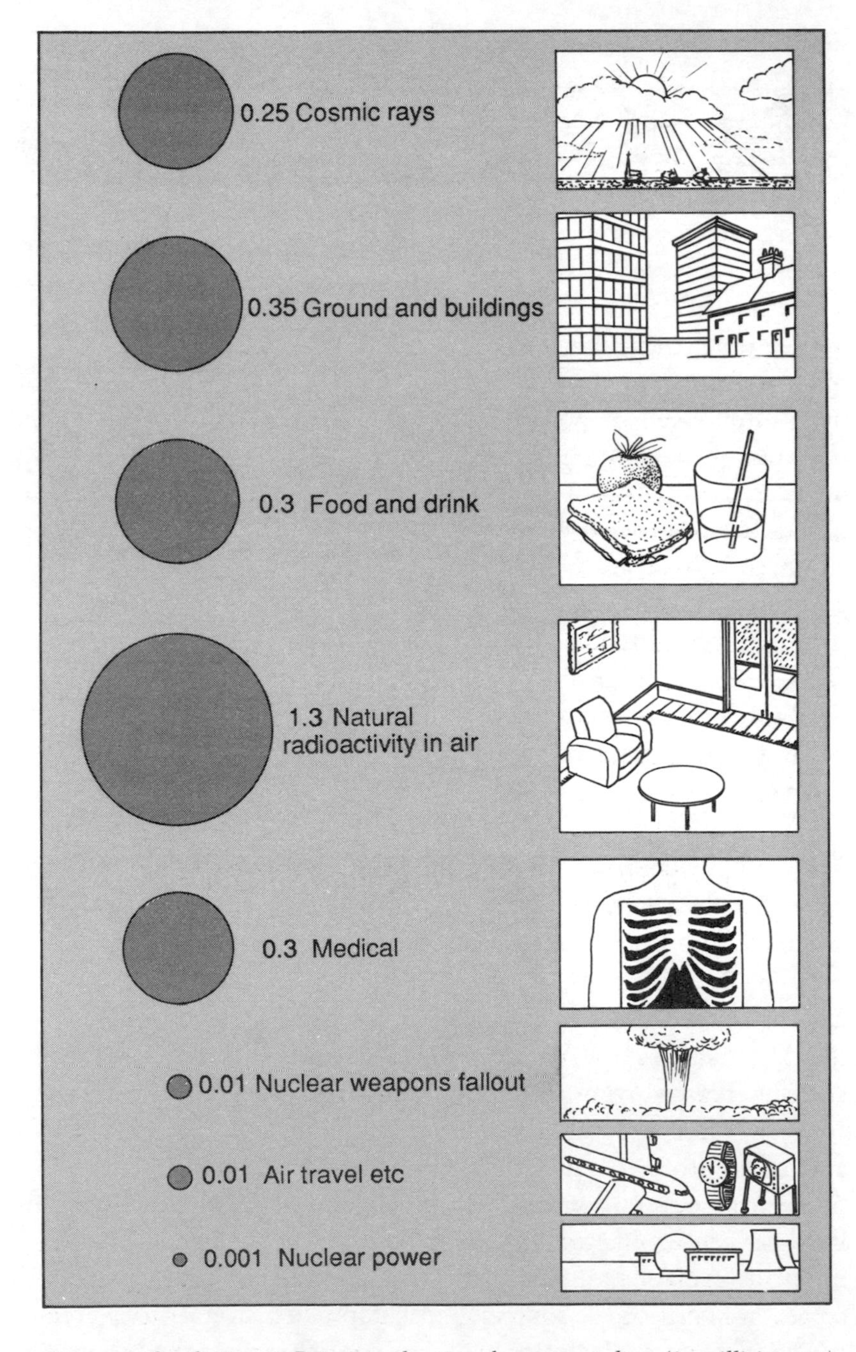

Sources of radiation in Britain – showing the average dose (in millisieverts) for each person

This source of radiation is particularly significant because we inhale radon and its decay products into our lungs. The sensitive bronchial tissues receive substantial exposures to alpha particles.

Cosmic radiation from outer space accounts for about 0.25 mSv per year on average. Someone living at an altitude of about 3000 metres, however, might receive a dose some three times greater. This is because at higher altitudes there is less atmosphere to absorb the cosmic rays before they travel through the body.

In normal circumstances man-made sources account for a small portion of the average dose of radiation. For example, nuclear weapons tested in the atmosphere, mainly in the 1950s, probably produce about 0.01 mSv a year. A similar dose results from air travel and miscellaneous sources such as TV sets and luminous watches.

The accident at Chernobyl in April 1986 added an extra 0.04 mSv – about the same as a return flight across the Atlantic. This figure is for the first 12 months after the accident; the level of radiation then rapidly declined.

It is the application of X-rays for medical purposes that accounts for by far the greatest proportion of man-made radiation – an average of 0.28 mSv a year, which is 94 per cent of the total due to medical procedures (6 per cent is due to nuclear medicine using isotopes). However, it is worth considering when *you* last had 'an X-ray' to realize that this particular contribution to the annual dose is far from evenly distributed among the population.

Today, we know that this radioactivity is the result of naturally unstable atoms changing from one variety into another. The inner parts of the atom rearrange themselves in an attempt to become stable. We also, however, deliberately create radioactive substances for purposes ranging from medical diagnosis to analysing the structure of materials. And we create radioactive by-products, particularly in nuclear reactors. We need to treat these artificial sources with care; radioactivity may be a natural process, but it can be dangerous, particularly when it is not properly controlled.

The atomic nucleus
Struggling for stability

Although radioactivity is a fundamental feature of the physical world, it was discovered only relatively recently. In 1896 a French physicist, Henri Becquerel, found that uranium salts could 'fog' photographic plates. (Becquerel made this discovery by accident, when he left some uranium salts in a drawer with photographic plates.) Some of the uranium atoms had changed, quite naturally, to atoms of slightly lighter elements. In so doing they had emitted radiation that left an imprint on the plates just as light does.

Physicists took another 40 years or so to form a reasonably complete picture of what radioactivity is. During this time they found that atoms are largely empty space, with tiny negatively charged electrons whirling around a compact, positively charged central nucleus. This nucleus consists of two types of particle – positive protons and electrically neutral neutrons, collectively called **nucleons**. Moreover, the number of electrons in an atom equals the number of protons, so that the atom is, overall, electrically neutral. The specific number of protons determines the element to which the atom belongs – one proton in hydrogen, two in helium and so on through to heavier nuclei such as uranium with 92 protons.

Although the number of protons in the nuclei of each element is always the same, the number of neutrons can vary slightly. Atoms with different numbers of neutrons are known as **isotopes** of an element. The presence of neutrons helped to solve the problem of why electrical repulsion between like-charged protons does not make the nuclei of atoms fly apart. Physicists invoked a force, called the strong nuclear force, or simply the **strong force**, which binds the nucleus together. This strong force exerts its influence over a very short range, not much farther than the distance across a proton – about a million millionth of a millimetre, or 10^{-15} metre (a femtometre). This force is about 100 times more powerful than the electric force at such short distances. It is also 'blind' to electric charge and acts equally on protons and neutrons, binding them together.

In relatively light nuclei, such as calcium with 20 protons and 20 neutrons, the strong force counteracts the electric force, provided the number of neutrons and protons is about equal. In heavy nuclei, however, there have to be more neutrons than protons to combat the repulsion between protons. The extra neutrons are there because the strong force acts mainly between neighbouring protons. The electric force, on the other hand, can act over longer distances. A proton at one side of the nucleus will feel some electric repulsion from a proton at the other side, but no strong-force attraction from protons or neutrons there. This means that the repulsive electric force begins to overcome the attraction of the strong force. The extra neutrons dilute the electric force and prevent the nucleus from blowing apart.

The balance between the forces is delicate. In particular, the subtle interplay of the strong and electric forces determines whether a particular nucleus will be stable or unstable. Some nuclei are rather like an inexperienced tightrope walker. They are unstable, and after some time they change towards a more stable configuration of protons and neutrons, just as the tightrope walker reverts to a more stable situation – the floor. It turns out with nuclei, however, that the route to stability can be as if via a succession of tightropes, some more stable than others.

A nucleus may make several transitions before it reaches stability. At each of these transitions, the nucleus must lose some energy, just as the tightrope walker loses potential energy (due to the height of the rope) in falling to the ground. The tightrope walker's potential energy may be converted to kinetic energy (energy of motion) as the person falls to the floor. Likewise, unstable nuclei emit excess energy, which we call radiation. Radioactivity is the process of making the transition towards a more stable state.

Radiation varieties
Steps in a decay chain

There are three main routes by which a nucleus can become more stable. Each results in a different type of radiation – **alpha particles**, **beta particles** or **gamma rays**.

Alpha particles are compact clusters of two protons and two neutrons that carry away the excess energy. They are, in fact, nuclei of helium. The helium nucleus is a very stable, tightly bound configuration of protons and neutrons. It is so stable that heavy nuclei, such as isotopes of uranium, find that it pays to rid themselves of two protons and two neutrons at once as they change, or transmute, towards a more stable structure. Materials that change in this way are called **alpha emitters** and occur mainly among the isotopes of elements heavier than bismuth. The stable form of bismuth – written as bismuth-209 or ^{209}Bi – contains 83 protons and 126 neutrons, or 209 nucleons in all. Notice that in emitting an alpha particle, a nucleus loses two protons as well as two neutrons. It therefore changes to the nucleus of a different element. Uranium-238, the commonest isotope of uranium, changes to thorium-234 in this way.

Sizing up radioactivity

There are three distinct units used by scientists to quantify radioactivity and radiation in the SI system.

The **becquerel**, named in honour of the physicist who discovered radioactivity, refers to the *activity* of a sample. One becquerel (Bq) is one nuclear transformation per second. It has replaced the curie, which was originally defined as the activity of 1 gram of radium, one of the radioactive elements that Marie Curie discovered. One curie (Ci) is equal to 3.7×10^{10} Bq.

Ionizing radiation loses energy in any matter it travels through, so we can quantify an 'amount' of radiation in terms of the energy it deposits, and thereby take into account the varying ionizing capabilities of the different kinds of radiation. In the SI system the unit called the **gray** is equivalent to one joule of energy being deposited in one kilogram of matter. It is a measure of *absorbed dose*, and in a sense reveals the intrinsic 'punch' of any kind of radiation. The gray (Gy) replaces the rad, with one gray being equal to 100 rads.

The amount of damage one gray of absorbed dose does to a human body depends on the type of radiation. Alpha particles, for instance, are inherently more damaging to biological tissues than electrons. This is because alpha particles lose their energy quickly by ionizing many atoms in a short distance, and can cause irreparable damage to cells. To quantify the potential of radiation to damage cells – and ultimately to kill – scientists use the **sievert**.

The sievert (Sv) is a measure of the so-called **dose-equivalent** – the absorbed dose modified by a quality factor that depends on the type of radiation. X-rays, gamma rays and electrons, for example, have a quality factor of one, whereas the quality factor for alpha particles is 20. Again the sievert has replaced an older unit, the rem (1 Sv = 100 rem).

There is no such thing as a completely safe level of radiation. A single electron, for example, could damage a cell irreversibly and initiate cancer. However, the likelihood of damage, and the severity of damage, increases with the amount of radiation.

An absorbed dose of less than 1.5 grays is extremely unlikely to lead to an early death, although the recipient may eventually die from cancer induced by radiation. A person receiving a dose of about eight grays, on the other hand, is almost certain to suffer irreparable damage to the bone marrow. The bone marrow is the source of all the cells that make up the immune system, and the person would probably die within two months from the consequent collapse of the immune system (the so-called haemopoietic syndrome). At still higher doses, of 10–12 grays or more, the radiation severely damages the intestine leading to death within a few days.

Beta particles, on the other hand, are energetic electrons. They are not, however, from the cloud of electrons that orbits the nucleus. Instead they are *created* by nuclei that have too many neutrons to be stable. These excess neutrons can turn into protons, emitting electrons as they do so, in a process called **beta decay**.

'Free' neutrons always change, typically after only 15 minutes, into protons which are slightly lighter in mass. The mass of the electron produced in the process accounts for some of this difference in mass, but it mainly appears as the kinetic energy of the electron and another particle that is created, the **neutrino**. Neutrinos have little, if any, mass and no electric charge. They interact only very weakly with matter: they can even pass right through the Earth without colliding with any of its vast numbers of electrons and nuclei. So the electrons constitute the only significant radiation from beta decays.

As in alpha decay, beta decay changes the nature of an element. In the case of beta radioactivity, the total number of nucleons remains the same, but the number of protons increases by one, while the number of neutrons decreases by one. Thus the element moves one step up in the periodic table of elements. Lead-210, produced in the decay chain of uranium-238, transmutes by beta decay to become bismuth-210, with 83 rather than 82 protons.

Beta decay of this kind arises in isotopes that are unstable because they have too many neutrons. Two other kinds of beta decay take place in nuclei with too many protons. Such nuclei do not occur naturally on Earth, but rather as a result of, for example, physicists bombarding a material with alpha particles. In these nuclei a proton turns into a neutron, and the nucleus becomes an isotope of the element one step down in the periodic table. One way for this transition to occur is for a proton in the nucleus to capture a nearby electron from the orbiting cloud. This process, called **electron capture**, creates a neutron and is accompanied by the emission of a neutrino. Electron capture tends to occur in heavier nuclei. Alternatively, a proton can change into a neutron and emit a neutrino

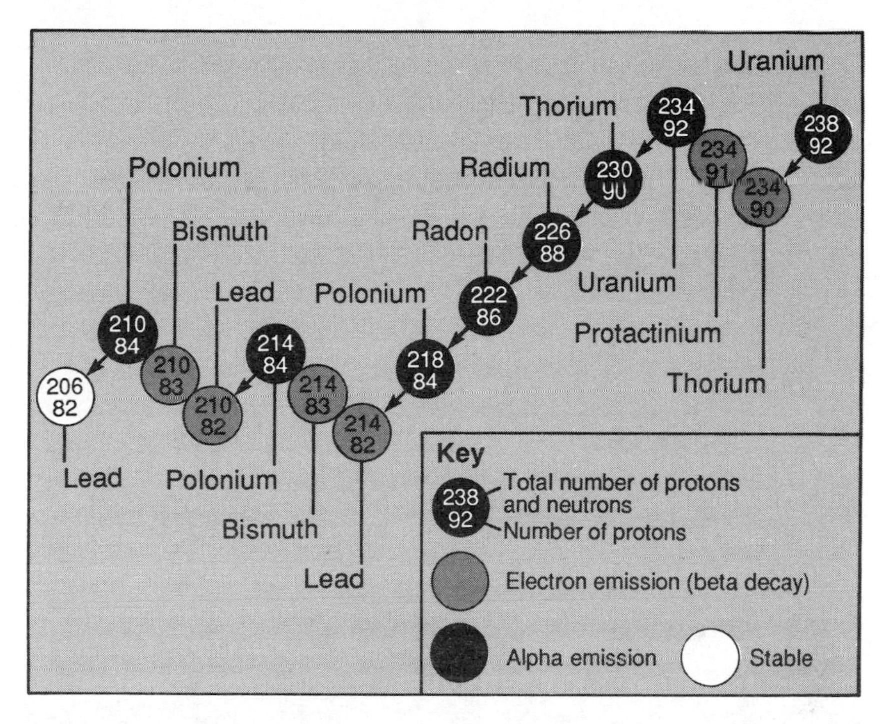

The main decay chain of the naturally occurring isotope, uranium-238. Ultimately, through a succession of alpha and beta decays, the radioactive nucleus reaches stability in the form of lead-206

together with a positron. A positron is an antielectron – it is similar to an electron except that it carries a positive charge. This type of beta decay is known as beta-plus decay, or positron emission.

The third main type of radioactivity gives rise to gamma rays. Gamma rays are electromagnetic radiation akin to light, but with very short wavelengths and very high energy. Often, when a nucleus forms through the alpha or beta decay of another nucleus, the new nucleus is 'excited' – it has more energy than is usual for this particular isotope. The nucleons rearrange themselves within the excited nucleus to lose the excess energy, emitting a gamma ray in the process. The number of protons and neutrons does not change when a nucleus emits gamma rays; the isotope retains its identity.

Atomic nuclei
Random decay

The rate of decay in unstable nuclei varies from fractions of seconds to billions of years. The timescale depends intimately on the balance of forces within a nucleus. But the underlying process is basically random. Physicists cannot predict at which instant a nucleus will decay; only the probability that it will decay within a certain time. For this reason, we talk of the **half-life** of an isotope. This is the time it takes for half the nuclei in a sample to decay. This means that after two half-lives, three quarters have decayed. And it takes seven times as long as the half-life to reduce a radioactive isotope to 1 per cent of its original quantity.

The decay pattern for all radioactive nuclei is the same, regardless of the half-life

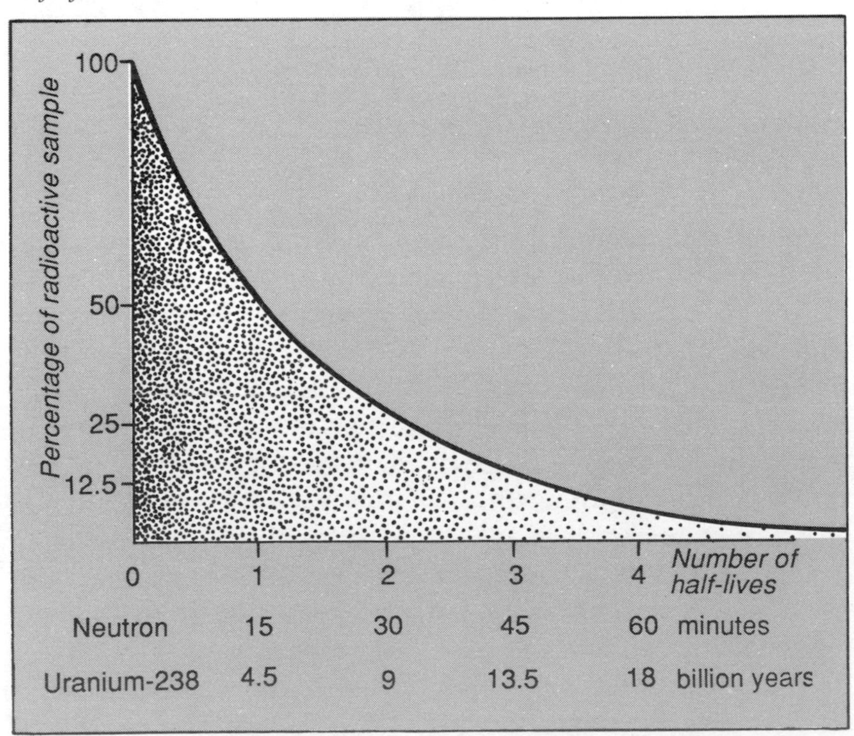

Some radioactive isotopes are long-lived, such as uranium-238 with a half-life of 4.47 billion years. This is roughly the age of the Earth, so only half the uranium-238 that existed when the Earth formed has decayed. Other isotopes, such as polonium-214, have half-lives measured in fractions of a second.

The long half-lives of certain artificially produced isotopes provide great problems in handling radioactive waste. For example, one by-product of nuclear reactors, neptunium-237, has a half-life of 2.2 million years. Any site for storing waste contaminated with such isotopes must be one that is unlikely to be disturbed in the future. Otherwise radioactive material may escape into the environment, with the attendant hazards.

Damaging effects
Medical weapons

It is the radiation that radioactive substances emit that has the potential to be harmful. The degree of harm to people depends on how they are exposed to radioactive materials, on whether for example they touch them or eat them as food; on the rate of radioactive decay in the material; and on the type of radiation emitted.

Alpha particles, beta particles (electrons) and gamma rays all damage materials in different ways. But the important point is that they all carry energy, which they lose in passing through matter. It is the transfer of this energy that causes the damage. The same is true for X-rays and cosmic rays – the rain of subatomic particles created in the atmosphere as high-energy protons, nuclei and gamma rays arrive from outer space. For this reason scientists, when discussing radiation and its effects on humans, usually include X-rays and cosmic rays as well as radiation due to radioactivity. The main way that radiation loses energy in materials is through ionization. Electrically charged particles, such as electrons, protons or alpha particles, interact with atoms they pass close to, and may knock an electron in such an atom out of orbit. These atoms, having lost negatively charged electrons, become positively charged – they become positive ions. Such changes give rise to radiation damage.

In biological tissue, ionization can lead to abnormal chemical reactions and molecular changes, which can destroy a cell or change how it functions. In particular, damage to the genetic material in a cell can lead to uncontrolled proliferation of the cell, which may result in cancer. Or, in reproductive cells, ionization may give rise to hereditary disease in the children of the individual. The degree to which ionization occurs depends on the speed at which the ionizing particle is travelling. The swifter the particle, the less it interacts with the atoms it passes and the smaller the amount of ionization over a given distance. In cosmic radiation, for example, particles called **muons** (a heavier version of the electron) are a relatively minor hazard. They are travelling close to the speed of light, and have little opportunity to deposit energy as they pass through our bodies.

The electric charge of the ionizing particle also affects ionization. For example, alpha particles have two protons bound to two neutrons and therefore two units of positive charge. They ionize matter four times more rapidly than single protons travelling at the same speed. This means that alpha particles lose their energy more rapidly than lone protons. The effects of charge and velocity together make the electrons from beta decay less ionizing than alpha particles, and they therefore travel farther through matter. A thin sheet of paper, or even the dead layer of skin on your fingers, is generally enough to stop alpha particles. Electrons from beta decay, however, travel farther, while still causing damage. It takes a thicker layer of material to stop them. Don't think that alpha particles are relatively harmless though: if you take them directly into your body (by eating or breathing), there is little to prevent them causing a great deal of damage.

Gamma rays and X-rays are not themselves electrically charged particles. They are forms of electromagnetic radiation and carry no electric charge. But they can still knock an

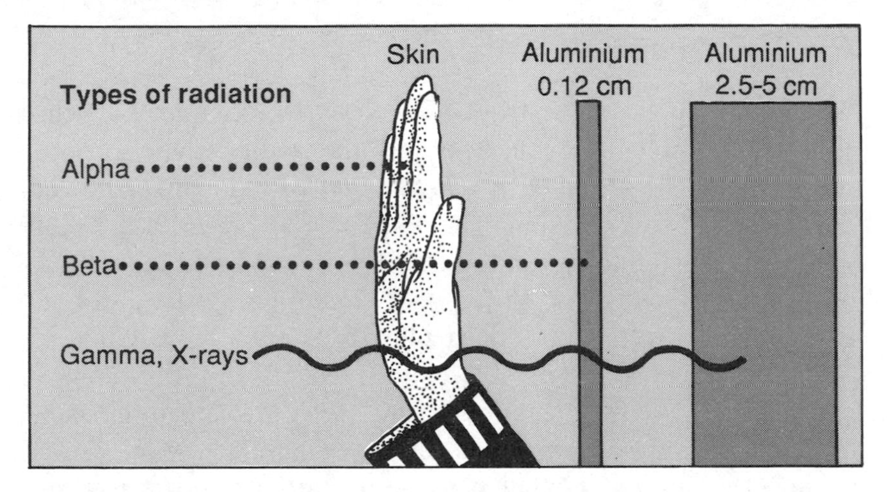

How far radiation travels through matter. Alpha particles lose their energy first – a layer of skin is enough to stop them. Beta particles are halted by a relatively thin sheet of aluminium. Gamma rays and X-rays are the most penetrating, and will pass through relatively thick layers before being absorbed

electron out of an atom or two. These electrons can then set about ionizing matter, causing damage in the same way as electrons from beta decay. Gamma rays are the most penetrating radiation to stem from radioactivity, and often the most hazardous. They will penetrate relatively great thicknesses of matter before they are absorbed.

Radiation and radioactivity, although potentially harmful, can also do good when properly controlled. Ever since Wilhelm Röntgen discovered X-rays in 1895, the medical profession has used them to reveal damage to bones or disease within the body. Doctors can also direct high doses of X-rays or gamma rays at tumours to destroy them. Another way to use radioactivity is to inject tiny amounts of radioactive elements into the body. These radioactive nuclei take part in biochemical reactions in the same way that normal nuclei do, and we can follow their radioactive emissions to locate tumours or other disorders, or to study oxygen metabolism or the brain. It is within our hands to use these natural phenomena properly, for our benefit.

11 February 1988

Further Reading

'Radioactivity', a poster from *New Scientist*, complements this chapter. It includes a chart of all the radioactive isotopes. *An Introduction to Radiation Protection* by Alan Martin and Samuel Harbison (Chapman & Hall, 1986) provides a useful guide to the terminology of radiation protection and the estimation of associated risks. *Living with Radiation* by the National Radiological Protection Board (HMSO, 1986) is a guide to radiation sources, doses, biological effects and radiological protection. Also by the National Radiological Protection Board is *Radiation Exposure of the UK Population: 1988 Review* (HMSO). Two books that provide a readable account of the history of modern physics, with chapters on the discovery of radioactivity and relevant areas of nuclear physics, are *From X-rays to Quarks* by Emilio Segré (Freeman, 1980) and *The Particle Explosion* by Frank Close, Michael Marten and Christine Sutton (Oxford University Press, 1987). *Radiation and Radioactivity on Earth and Beyond* by Ivan and Zorica Draganić and Jean-Pierre Adloff (CRC Press, 1990) is a very clear introduction to the many aspects of radioactivity.

CHAPTER 9

Rocky Dwarfs and Gassy Giants

Nigel Henbest

Only 30 years ago astronomers knew little about the other planets, the Earth's neighbours in space. So far as they were concerned, nine planets orbited the Sun; the four closest to the Sun were small, the next four much larger, and Pluto – at the edge of the Solar System – another small planet. Even with today's telescopes, astronomers on the Earth can tell relatively little about the other planets. For a start the other planets are a very long way off; even the nearest, Venus, never comes closer than 38 million kilometres. When the light from one of the planets penetrates our atmosphere, moving currents of air distort our view. In the late 1950s engineers developed rockets powerful enough to take spacecraft well beyond the Earth and Moon, and off to the other planets. Space probes from Earth

Apart from tiny Pluto, we can divide the planets of the Solar System into two types. Close to the Sun are the four 'dwarf' planets, which are dense because they are made of rock and iron. The next four are 'giants' made of gas. Saturn has such a low density that it would float in water, if we could find an ocean big enough!

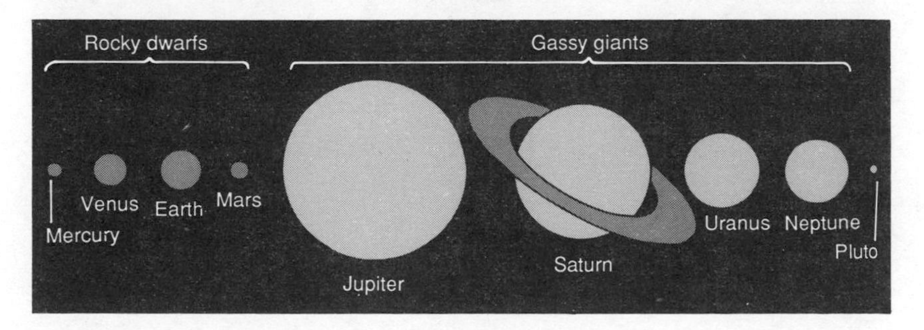

have now flown past all the planets except Pluto, and have also photographed a comet at close quarters. Some craft have landed on Mars and Venus.

Even before the advent of space probes, astronomers had some idea of what the planets were made of. They could see through their telescopes how big a planet was. They could work out its mass – from the pull of its gravity on moons orbiting it, or, less accurately, from its pull on other planets. From such information the astronomers could calculate its density. The densities of Mercury, Venus and Mars are very similar to the Earth's, so they too must be made largely of rock (possibly with a core of iron). Astronomers named these planets the **rocky dwarfs**. The next four planets are obviously different; they are all very much larger than the Earth, and their densities are much lower – not much more than that of water. Astronomers realized that the low density meant that these planets consist mainly of gases or liquids; so they called them the **gas giants**. At the edge of the Solar System is tiny Pluto. Because no spacecraft has yet visited it we still know very little about it, but astronomers believe that it is a small solid planet made of a mixture of rock and ice.

Patterns of weather
Out of this world

As we all know, predicting the weather is a thoroughly difficult business. This is because the Earth's atmosphere behaves in a complicated way, even though the basic processes that drive the weather are simple. The Sun heats the equator more than the poles, so the air at the equator rises, moves north or south, and comes down at the poles, only to travel back to the equator at a lower level. The Earth's rapid rotation twists these wind flows around, so there are also east-west and west-east winds at different latitudes.

By studying other planets, in particular Venus and Mars, meteorologists have been able to test their theories, and improve

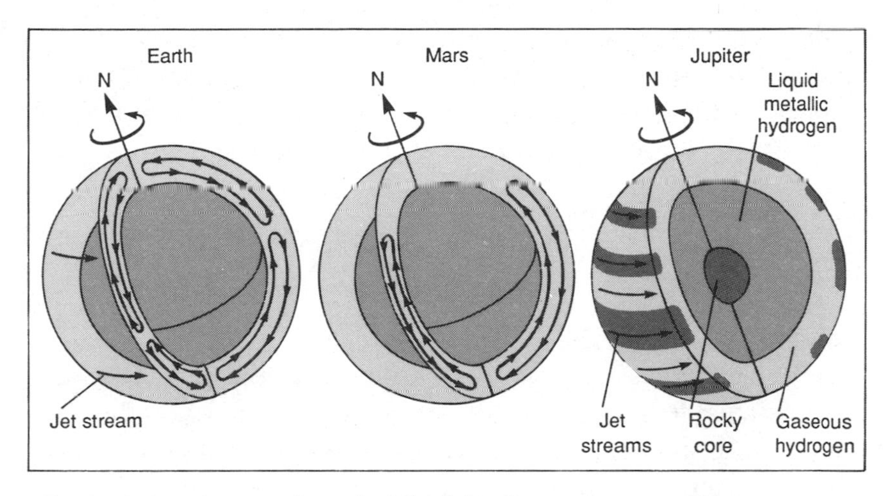

The Sun's heat (coming from the left of this diagram) causes different weather patterns on the planets. On Earth, hot air rises at the equator and moves to the poles, where it cools and goes down. On Mars, hot air rises in the hemisphere that is summer, and comes down in the winter hemisphere. On Jupiter, swirling eddies of gas produce 'jet streams' that speed around the planet: eddies in the Earth's air produce less powerful jet streams

them. Venus, for example, rotates slowly – once in 243 days – but winds at the top of the atmosphere move extremely fast, up to 300 kilometres per hour. On Mars, regions of high and low pressure (anticyclones and depressions) exist which are similar to those that we all experience from time to time on Earth. But on Mars they only occur in one hemisphere at a time, in the half of the planet that is then experiencing winter. (The seasons there last almost twice as long as they do on Earth.)

The weather on the gas giants is surprisingly similar to ours; surprisingly, because our weather occurs only in a thin layer of gas over a solid surface, whereas the gas giants consist of fluid (gas or liquid) throughout. Meteorologists believe that these giants have weather similar to that on Earth because the same processes operate on them. The Sun's heat (and heat from within, in the case of Jupiter, Saturn and Neptune) causes small swirling eddies of gas to rise. These eddies give their energy to winds that blow in various directions at different latitudes. What has surprised scientists is that this process is much more efficient on the gas giants. In our atmosphere, eddies convert one thousandth of the total energy in the atmosphere into high-speed winds, known as jet streams. Jupiter's eddies turn one tenth of the energy in the atmosphere into winds. Neptune seems to be even more efficient.

Magnetic fields abound
Age-old faces

The gas giants can tell us a lot about how the planets create magnetic fields. Of the four dwarf planets the Earth is the only one with a strong magnetic field; this it generates by electric currents deep within its molten iron–nickel core (see Chapter 11). Both Jupiter and Saturn have magnetic fields stronger than the Earth's. These rapidly spinning giant worlds consist largely of hydrogen, but it is under such pressure that it behaves rather like a liquid metal and conducts electricity. So although these planets have hydrogen 'metal' rather than iron at their core, they seem to confirm the basic theory of the Earth's magnetism.

The next two planets, Uranus and Neptune, both produce a magnetic field, but not in their core. On Neptune the magnetic field is generated within a region four fifths of the way from its centre to the surface. In both giants the magnetic poles are not near the planet's real poles (the ends of its axis of rotation) but lie half-way to the equator. Astronomers think that here the region of electrically conducting fluid lies above a large, insulating core. This idea ties in with theories that Uranus and Neptune have cores made of rock, surrounded by a thick layer of water. Water that contains impurities is a good conductor of electricity, so on these planets the magnetic field originates in a deep ocean of warm water.

Since the birth of the Solar System the Earth has altered because geological processes have changed its surface. Our Moon and Mercury are two worlds where there has never been much geological activity. Their surfaces are good museums of the way that all the rocky planets must have looked soon after their birth. Both the Moon and Mercury are pocked with craters; giant pits blasted out by the impact of solid bodies falling in from space.

The manned Apollo flights to the Moon brought back a third of a tonne of Moon-rocks. When scientists had measured the ages of these rocks, they were able to say that most of the Moon's surface has remained unchanged for almost 4 billion years, so it is indeed a museum of the early days of the Solar System. Just as important the Apollo results showed that scientists could tell the age of a region of the Moon from the number of craters on it. A once pock-marked area may have disappeared beneath lava flows, and show only few craters as there are now very few solid bodies whizzing around in space to crash into it. By counting craters, astronomers can tell, roughly at least, the ages of portions of the other planets.

The first spacecraft to Mars disappointed astronomers who had hoped to find a planet similar to the Earth, perhaps one that sustained life. The pictures from Mariner 4 in 1965 and Mariners 6 and 7 in 1969 showed a planet covered in craters almost as thoroughly as the Moon. We now know that we were unlucky with our first close-up views of Mars. In 1971 Mariner 9 went into orbit around the planet, and found that previous spacecraft had shown only half the picture – quite literally. Whereas half of Mars consists of a heavily cratered surface, the rest is entirely different. Here, we find vast volcanoes and canyons 3000 metres deep, both larger than anything on the Earth, and signs that water once flowed on the now frozen surface. These results – combined with those of Vikings 1 and 2 in 1976 and the Soviet Phobos-2 in 1989 – show that 'geological activity' has been happening on Mars, and quite recently in terms of the age of the Solar System. Flows of lava and erosion by running water have erased the older craters on more than half of the surface of Mars.

The largest volcano, Olympus Mons (Mount Olympus), is 25 kilometres high – nearly three times as high as our own Mount

Everest – and it has very few craters caused by impacts from space. This scarcity of craters means that Olympus Mons is young in geological terms, and eruptions would have occurred until about 100 million years ago. In fact, this volcano may merely be 'dormant', and may erupt again in the future.

Geology of a twin
Plateaux and plains

Astronomers and geologists hoped that Venus would provide some clues to an important question: Would any planet that is similar in size to the Earth necessarily have the same kind of geology? The problem is that Venus is covered by cloud. In recent years, however, astronomers have built up a picture of the surface below the cloud, using radar. With this technique they send radio waves through the clouds, and reconstruct an image of the surface below from the way in which the waves are reflected by the rocks. Venus has two large, high plateaux and several volcanoes. The rest of the planet is covered by low-lying plains. The most recent radar results show very few impact craters on these areas, indicating that they are young; no more than a few hundred million years old. In fact, lava probably seeps out on to these plains even now.

Although both the Earth and Venus have volcanoes and lava plains, the geology of the two worlds differs in one important respect. The Earth's surface is broken up into a couple of dozen plates, which move around the Earth's surface by a process called **plate tectonics** (see Chapter 10). Where two plates collide earthquakes happen, or a mountain chain (such as the Himalayas), even a line of volcanoes, may form. But so far, images of Venus have not indicated the same kind of plate tectonics. Recent radar pictures show that Venus's surface is probably splitting around the equator, and that the plates spreading north and south are crumpling into folds as they move.

Strange activities
Freezing eruptions

One great surprise from the Voyager missions to the giant planets was that most of their moons showed signs of geological activity having erased the old cratered surface. Although these moons look rather like our Moon, or like parts of Mars, there is an important difference. At more than 700 million kilometres from the Sun, water is frozen solid into ice – and, at the distance of Uranus and Neptune, ammonia and methane are frozen too. So their moons are made up of a mixture of rock,

Sizing up the planets

	Average distance from Sun (million km)	*Equator diameter (km)*	*Mass relative to Earth*	*Average density relative to water*	*Number of satellites*	*Names and diameter of main satellites (km)*
Mercury	57.9	4878	0.055	5.42	—	
Venus	108.2	12 103	0.81	5.25	—	
Earth	149.6	12 756	1.0	5.52	1	Moon (3476)
Moon	—	3476	0.012	3.34	—	
Mars	227.9	6794	0.11	3.94	2	Phobos (22) Deimos (12)
Jupiter	778.3	142 800	318	1.32	16	Ganymede (5262) Callisto (4800) Europa (3138) Io (3630)
Saturn	1427.0	120 660	95	0.69	18	Titan (5150) Rhea (1530)
Uranus	2869.6	51 400	15	1.26	15	Titania (1590) Oberon (1560)
Neptune	4496.7	49 400	17	1.64	8	Triton (2720)
Pluto	5900	2280	0.002	2.1	1	Charon (1200)

ice and perhaps other frozen material. The fact that these worlds consist of different materials from the inner planets is very useful. It shows what kinds of general geological process can occur, whatever the world is made of; and the differences that depend on its particular make-up.

Saturn's moon Enceladus, for example, has smooth plains that probably consist of ice that flowed from its interior as water 'lava' when part of the moon's interior melted. The surface of Europa, a moon circling Jupiter, is covered with a smooth, brilliant-white plain. This is probably the surface of an ocean of water that once covered the moon, but froze solid to make it look like a billiard ball. Triton, the largest moon of Neptune, also has few impact craters, indicating that it, too, melted completely at one time. But its surface is not smooth. Triton is covered by polygonal rings, long low ridges and frozen lakes of ice. When the molten Triton cooled down, its surface did not simply freeze – as did Europa's. It seems to have experienced some complicated 'geology' of a type that scientists do not yet understand. Even stranger is the surface of Uranus's moon Miranda. Here we find large V-shaped features and oval rings of ridges and troughs. So far, no one is sure how they formed. Some scientists think this moon once broke up into half a dozen pieces, which then reassembled. But it is more likely that the strange markings are regions where part of the moon's interior softened and churned up the surface.

As Voyager 1 passed Jupiter's moon Io in 1979, its cameras revealed umbrella-shaped plumes of gas rising 300 kilometres into space. Ten years later, Voyager 2 found a smoke-like eruption of dark material from Triton, Neptune's largest moon. Although people often refer to these outbursts as 'volcanoes', they are more similar to geysers on the Earth. The difference is that a volcano spews out the actual hot material that rises from inside the Earth – molten lava, or **magma**. In a geyser, the magma is trapped below ground and its heat boils water in the rocks above to make the water erupt. Io has molten rocky magma below the surface, and it heats up sulphur or sulphur dioxide which then erupts through the surface as a plume of vapour.

Triton is much farther from the Sun, and the temperatures are much lower. Indeed, at −236 °C – only 37 degrees above absolute zero – the surface of Triton is the coldest place we know in the Solar System. Here the 'hot' material inside is ordinary ice, trapped deep within Triton. It heats up pockets of frozen nitrogen, which burst through the surface as nitrogen gas, carrying small specks of black, probably organic, material from below and spreading them into dark streaks seen by Voyager's cameras. (Our experience from close-ups of Halley's Comet tells us that 'black material' in the Solar System is usually associated with carbon.)

A long family story
Birth of the planets

Geological processes such as erosion and plate tectonics have destroyed or hidden the earliest rocks on the Earth, so we cannot tell directly how our planet was born. However, astronomers can now piece together the early history of the Earth and the other planets from their studies of the other worlds – including the rocks brought back from the Moon and meteorites that fall to the Earth from space.

About 4500 million years ago, a cloud of gas and dust in space collapsed under its own gravity to form a dense blob at the centre, which became the Sun, and a swirling disc of matter farther out. The dust and gas in this disc would eventually become the planets that we know today. The solid 'dust' in the disc consisted of particles of ice and rock, only a millionth of a metre across. In the regions closest to the centre the heat of the nascent Sun boiled away most of the icy particles, to leave mainly grains of rock. That is why the planets near the Sun are rocky. Towards the edge of the Solar System, where it was colder, the icy particles survived, and we find the planets Uranus and Neptune made up of a mixture of rock and ice (now melted). Jupiter and Saturn began in the same way but became so massive that their gravity was able to capture some

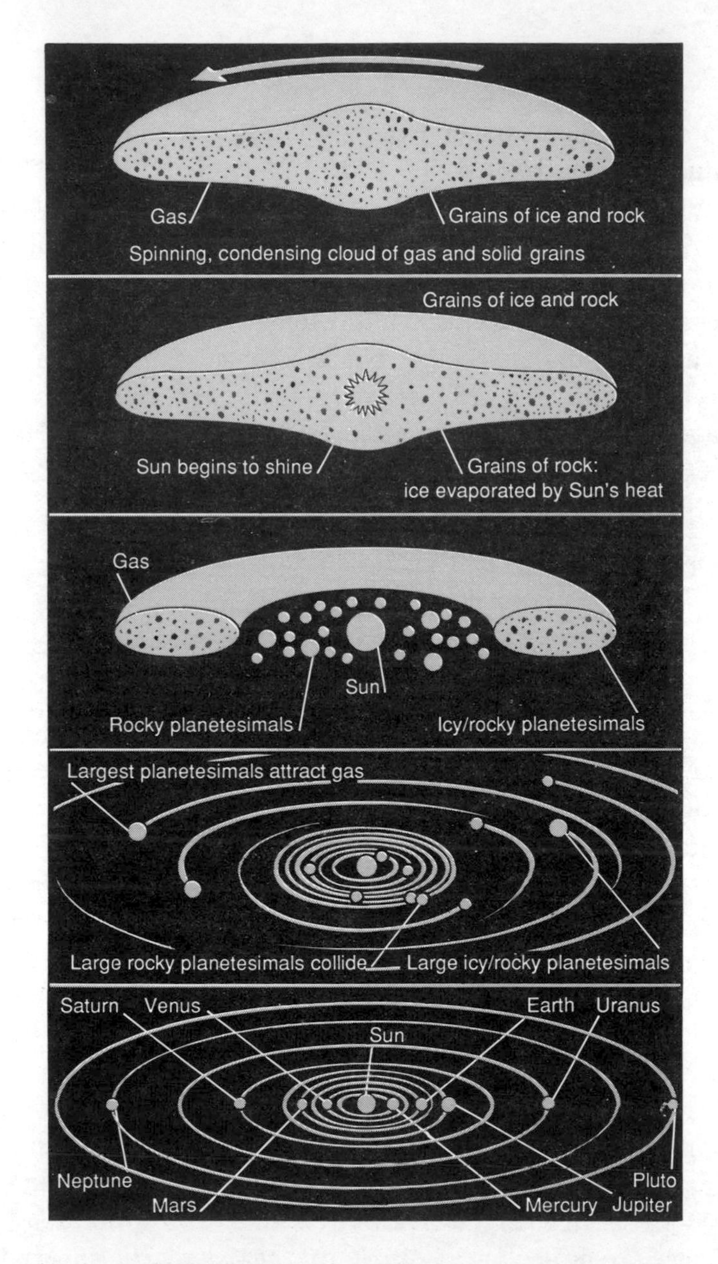

The Solar System – the Sun and the nine planets – started off as a spinning cloud composed of gas and solid grains of ice and rock. When the Sun formed in the centre, its heat evaporated the icy grains in the inner part of the Solar System, leaving only grains of rock that grew into the four 'rocky dwarfs'. Further out, the icy particles survived, so the planets that formed here were larger: their gravity then attracted extra gas, to build them into 'gas giants'. The grains first clumped into planetesimals, a few kilometres across, which subsequently collided to build up the planets

of the hydrogen and helium in the disc, so they now consist mainly of these gases.

Astronomers think the way that the planets came together was much the same, regardless of whether the original particles were icy or rocky. In the dense disc, these 'dust' particles merged to form **planetesimals**, a few kilometres across. How the particles accumulated into planetesimals, we do not yet know; they may simply have stuck together when they collided, or possibly the gravitational effect of all the particles on one another became unstable, and the matter in the disc broke up into clumps which then coagulated.

The theory becomes firmer after the planetesimals formed. Astronomers have calculated by computer how planetesimals would collide with one another, sometimes crashing at such speed that they break one another up, sometimes meeting at a lower speed so that gravity binds them together. Any clump of planetesimals that grows to be bigger than the others will have a stronger gravity, and so will be better at holding on to other bodies that it collides with. As a result, the largest clumps of planetesimals tend to gather up the others and become the cores of the planets. These calculations show that, in the inner part of the Solar System, the planets formed from planetesimals in about 200 million years; the outer planets, which formed in regions where there was less matter about, took rather longer.

The final stages of forming the planets probably involved high-speed collisions between quite large bodies, and with very dramatic effects. Scientists claim that some calculations suggest Mercury originally formed farther out than Mars, but a tremendous collision drove it in towards the Sun, stripping it of its outer layers of rocks. The composition of lunar rocks suggests that the Moon was born from a similar collision, now nicknamed the Big Splash. When the Earth had reached almost its present size, another body about the size of Mars crashed into it. The 'splash' threw out a spray of molten rock, which condensed into a ring of particles surrounding the Earth. These rocky particles then came together to make up the Moon.

After the basic nine planets had formed, with their attendant satellites, they quickly swept up most of the planetesimals. The impact of these smaller bodies left the craters we see on all the old surfaces that have survived in the Solar System. Some planetesimals still exist. The asteroid belt, beyond Mars, consists of rocky planetesimals that never accumulated into a planet because Jupiter's gravity kept shuffling their orbits.

Running rings around the planets

Even a small telescope reveals that Saturn has a set of rings. Astronomers deduced that Uranus also has rings when, in 1977, the planet moved in front of a star and the star's light was cut off briefly just before and just after the planet itself hid it. Similar observations of Neptune passing in front of stars led some astronomers to believe that it too had rings.

The Voyager spacecraft confirmed that there are rings around Uranus and Neptune, and discovered rings around Jupiter as well. They also showed the rings of Saturn in great detail.

Astronomers were surprised to find that the rings of Uranus and Neptune are very narrow. They are only a few kilometres thick, even though they orbit the planets at a distance of around 50 000 kilometres. (On a scale model, the rings, given a radius of about one metre, would be as thin as strands of thread.) The Voyagers found that even the broad rings of Saturn are made up of thousands of narrow 'ringlets', nested one inside the other.

These rings are composed of millions of chunks of ice, ranging in size from a few millimetres up to several metres, and each of them orbits the planet as if it were a tiny moon. The gravitational forces of small moons that orbit the planet just inside and just outside each ring seem to 'shepherd' chunks of ice, keeping them in narrow rings.

Most important to astronomers, the chunks in these rings must be moving in a similar way to the pieces of rock that made up the Solar System in its earliest days. The rings of the planets, therefore, can help astronomers in their efforts to understand how the planets came together from smaller pieces of rock and ice.

Beyond the planets there is a region where we find large numbers of icy planetesimals, each a few kilometres across. These become comets when their paths are altered so that they come nearer the Sun. Eventually, solar heat causes them to evaporate, which gives them long gaseous tails.

Now that astronomers have made their first reconnaissance of the major planets and the large moons, they are turning their attention to the smaller worlds that are fragments left over from the earliest days of the Solar System: the asteroids and comets. Late in 1989 astronomers obtained their first view of the surface of an asteroid, 1989PB, by bouncing radar waves off its surface. They found it to be shaped like a peanut. The spacecraft Galileo, now on its way to Jupiter, will take the first close-up pictures of two asteroids, Gaspra in 1991, and Ida in 1993. Astronomers are also planning a spacecraft to take a long look at a comet; the first comet missions went to Halley's Comet in 1986, but they gained only a fleeting glimpse because they passed it at 240 000 kilometres an hour. A detailed study of the surviving planetesimals should fill in the missing details about how the major planets – including the Earth – came into being.

10 February 1990

Further Reading

The Cambridge Photographic Atlas of the Planets by Geoffrey Briggs and Fredric Taylor (Cambridge University Press, 1986) has photographs and some detailed maps of the planets out to Uranus, while *The New Solar System* edited by J. Kelly Beatty, Brian O'Leary and Andrew Chaikin (Cambridge University Press, 1983) goes into the science in more depth. *The Planets* by Heather Couper and Nigel Henbest (Pan, 1986) provides a colourful introduction to the Solar System; it includes a loose-leaf insert on spacecraft observations of Uranus and Halley's Comet in 1986.

PART 2

Earth Sciences and the Environment

CHAPTER 10

Structure of the Earth

Richard Fifield

Squeeze yourself and you can say with certainty that flesh is more substantial than air or water, but less so than the rocks we tread. The density of water is 1 gram per cubic centimetre (g/cm^3); the average density of the Earth is 5.5 g/cm^3. The general density of our bodies, even allowing for bones, is much closer to 1 than it is to 5.5 g/cm^3.

A sense of density also proves to be a useful way of finding out about the structure of the Earth. We cannot, of course, look directly inside our planet; light cannot travel through rock. Pressure or shock waves from earthquakes, however, can. As these waves pass through rock, their speed and course change as the rocks change. Geologists can, therefore, probe the rocks deep in the Earth's interior by measuring how long it takes shock waves to travel from an earthquake to various points around the globe. What this reveals is that the Earth is made up of layers, rather like an onion. On the outside is a thin **crust** – its depth no more than a postage stamp stuck on the outside of a football. Underlying this is a **mantle**, which makes up more than 82 per cent of the volume of the Earth. Deeper still, we come to a very dense and very hot **core**.

During this century geologists have pieced together numerous bits of information to form an increasingly detailed picture of the interior of the Earth. Much of this information comes from studying earthquakes. But geologists have also turned to the Earth's magnetic field, meteorites, chains of islands and volcanoes. All reveal something about the inside of our planet.

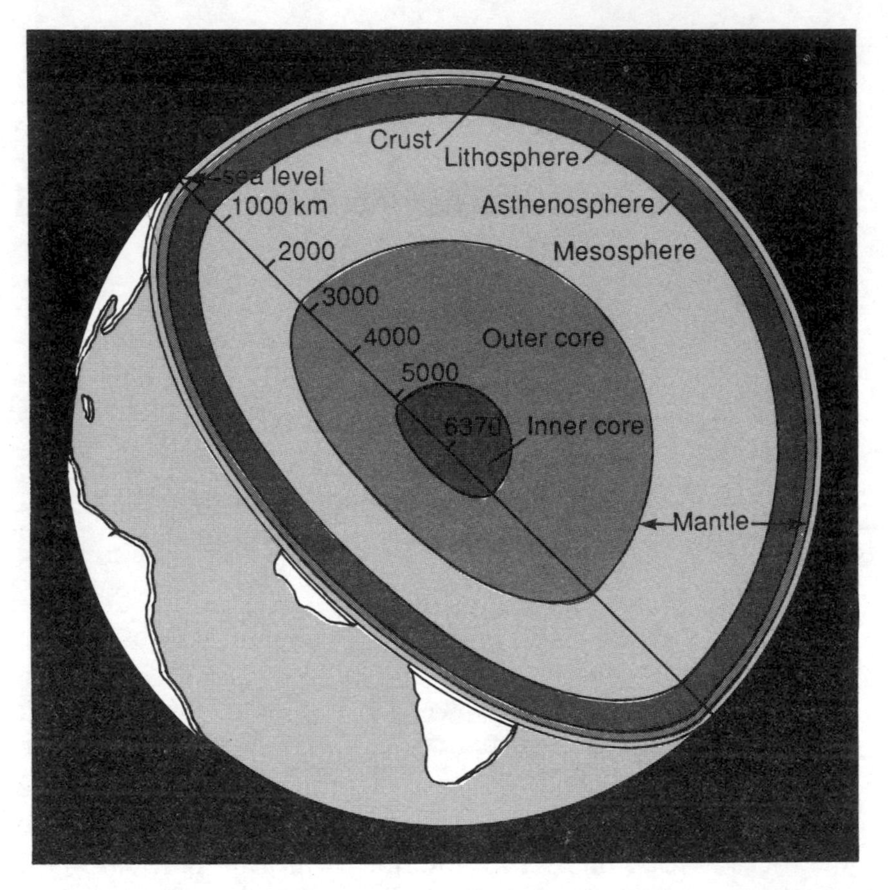

The layers of the Earth. The crust on which we live represents a mere 1 per cent of the volume of the Earth, and is part of the rigid outer shell, the lithosphere

Layer by layer
The planetary onion

After an earthquake, pressure or shock waves spread out in all directions. Like light waves passing through glass, these shock waves are reflected or refracted when they meet a rock with a different density. If the rock that the waves pass through becomes denser, they speed up; if the rock becomes less dense, they slow down. By determining the path and speed of these seismic waves through the Earth, geologists can identify the density and thickness of rocks that lie thousands of kilometres below our feet.

Scientists developed the first instruments, called **seismographs**, for recording shock waves at the end of the last century. They soon established that an earthquake creates two main types of wave in the interior of the Earth. The first, the **primary**, or **pressure (P)**, waves, travel by alternate compression and expansion. Such waves can pass through rocks, gases and liquids. The P wave is followed by the **shear**, or **secondary (S)**, wave. Because the S wave travels with a side-to-side vibration, it can pass only through solids (liquids and gases have no rigidity to support the sideways motion).

When Andrija Mohorovičić, a Yugoslavian geophysicist, analysed the records of an earthquake in Croatia in 1909, he detected four kinds of seismic pulse, two of them pressure and two shear. Seismographs close to the site of the earthquake recorded slow-travelling S and P waves. In recordings made farther away from the earthquake, these signals soon died out and were replaced by faster S and P waves. Mohorovičić interpreted the slow waves as ones that had travelled from the focus of the earthquake to the seismograph station directly through the upper layer of the crust. The faster waves, however, must have passed through an underlying layer of denser rocks, which deflected them and increased their velocity. He concluded that a change in density from 2.9 to 3.3 g/cm^3 marked the boundary between the Earth's crust and the mantle. In recognition of his discovery geologists call this boundary the **Mohorovičić discontinuity** or, simply, the **Moho**.

As seismologists accumulated records from seismographs they detected a 'shadow' zone between 105° and 142° from the source of the earthquake where they could not detect the shock waves. Beyond 142° the P waves reappeared on seismic records. The only explanation that fitted was that the shock waves had passed from a solid to a liquid. This would stop the S waves, and refract and slow the P waves. The seismologists concluded that the density changes from 5.5 to 10 g/cm^3 at a depth of

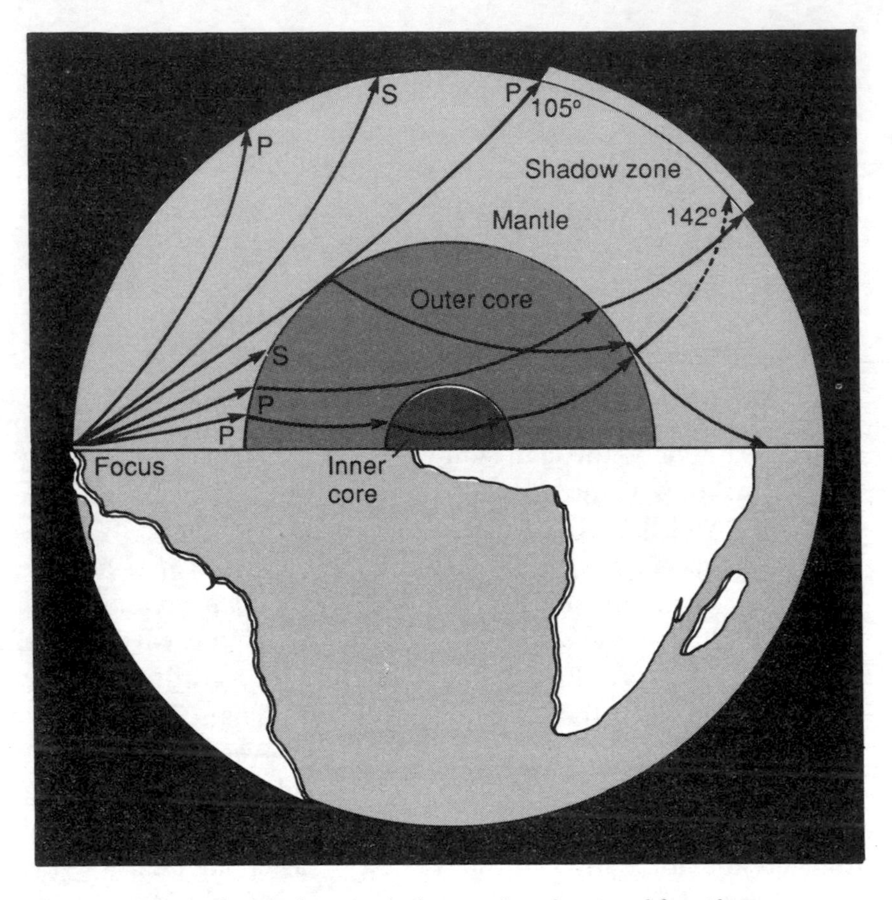

Seismographs in the 'shadow zone' of an earthquake record few, if any, shock waves. This led to the discovery of the core

2900 kilometres. They designated this the boundary between the mantle and core.

Later, however, they detected some waves, albeit much fainter, in the shadow region. In the mid-1930s a Danish seismologist, Inge Lehmann, suggested that a further change in density occurred at about 2250 kilometres into the core. This density change would accelerate the P waves, and bend some of them so that they emerged in the shadow region. She concluded that inside the Earth there is an inner core of a very dense solid. Others confirmed her conclusions, and we now estimate that the density changes from 12.3 to about 13.3 g/cm^3 at this boundary, before reaching 13.6 g/cm^3 at the centre of the Earth. The **outer core** is, then, a fluid metal and the **inner core**, solid metal.

Temperature *v.* pressure Solid to plastic

Increasingly nowadays, geologists probe the interior of the Earth with artificial seismic waves – using echo sounders, small explosions and other percussive systems which they can turn on and off where and when they wish. The overall picture we now have is of a series of concentric layers, which become progressively denser towards the centre. Two opposing factors control the rigidity, and hence density, of the layers. The first is temperature, which works to soften or melt rocks. Much of the inner Earth is white-hot from the energy produced by the decay of radioactive elements in the rocks. At the centre of the Earth the temperature is perhaps 3000 °C, falling to 375 °C at the mantle–crust boundary. The second factor is pressure, which tends to solidify rocks. The deeper you go, the greater the weight of overlying rocks, and the higher the pressure.

In effect, close to the cool surface, rocks are mainly solid and brittle. Geologists call this region the **lithosphere** (from the Greek *lithos*, stone). This layer, which takes in the crust and the upper mantle, is about 70 kilometres deep. At this point, seismic waves slow down indicating a change of density. This is the **asthenosphere** (*asthenos* is the Greek word for 'lack of strength', that is to say, plastic). Here, radioactively produced heat cannot dissipate so readily, and the rocks tend to melt and may even flow. The 'toffee'-like asthenosphere extends for almost 200 kilometres. Below the asthenosphere the seismic waves at first accelerate rapidly, and then slowly increase in speed for about 2500 kilometres. This is the **mesosphere** (from the Greek *mesos*, middle). Despite the heat, the effects of pressure are so great that rock and materials take on a rigidity and can 'creep' only very slowly. At the core the S waves peter out. The temperature at first is high enough to counteract the immense pressures from the rocks above, and for a depth of 2200 kilometres the core is liquid. Finally, the pressure builds up until in the centre there is 1270 kilometres of solid, inner core.

Compositional conjectures
Meteorite clues

It is one thing to know something of the physical state of material in each of the inner zones of the Earth, but what are they made of? Although we can drill boreholes and remove samples of rock, no one has reached the mantle yet. The world's deepest hole, in the Kola Peninsula in the Soviet Union, penetrates only 12.6 km into the interior – about half-way through the crust, or 0.2 per cent of the way to the centre of the Earth. So what do we know of the crust by direct observation? In continental areas we find that silicon and aluminium are abundant. These elements, combined with oxygen, make up the most common rock – **granite**. Below the oceans, and underlying the continental granite, the crust mainly consists of **basalt** rock, in which silicon, iron and magnesium predominate.

There our certainty ends – though geologists believe they have discovered rocks that have made their way up to the surface from the mantle at four sites: in north Italy, south-east Turkey, the Persian Gulf and New Guinea. These dark, heavy rocks, known as **peridotites**, are composed of olivine and pyroxene – silicate minerals which are formed only at high pressure and are rich in iron and magnesium. The density of the peridotites is such that shock waves from earthquakes would travel in them at similar velocities to seismic waves in the mantle. The best guess at present, therefore, is that the mantle largely consists of oxygen, silicon, magnesium and iron. In the upper mantle these elements probably exist, in mineral form, as olivine, pyroxene and garnet. As the pressure from overlying rocks increases, the atoms would rearrange themselves in more compact forms, as **high-pressure minerals**, and this in turn would change the structure of the rock. Ultimately, in the lower mantle, the minerals would probably break down into simple oxides.

Meteorites give us other valuable clues to our planet's composition. Most of the meteorites that reach the Earth's surface are one of two types – either stony or metallic. If, as scientists believe, meteorites represent the remains of planetary bodies similar to the Earth, then the stony meteorites are likely to represent mantle-type material, and the metallic ones pieces of a former core. If this idea is right – and it seems that the composition of the stony meteorites is broadly similar to what we know of the mantle – we can learn much about the core of the Earth by studying the metallic meteorites. They contain predominantly iron, iron sulphide and a range of the so-called siderophile elements, including nickel, platinum and other trace metals such as iridium. It is the metals in the core that give rise to magnetism.

A magnetic field
Natural dynamo

The Earth has a strong magnetic field, the origin of which puzzled scientists throughout the ages. Although the ancient Chinese and Greeks seem to have known something about the planet's magnetic field, it was not until the seventeenth century that we understood the nature of the field. An English physicist and physician, William Gilbert, suggested that the Earth acts as an enormous bar magnet. His theory held good until this century. Then scientists detected that at roughly every 200 000 to 300 000 years, the Earth's magnetic field reverses: north becomes south and south becomes north. Such **reversals** disagree with the idea of a permanent bar magnet and its field.

In the past 30 years or so an explanation has emerged. As the Earth spins on its axis, the fluid layer of the outer core allows the mantle and solid crust to rotate faster than the solid inner core. As a consequence, claim some geophysicists, electrons in the core move relative to those in the mantle and crust. The movement of the electrons effectively establishes a natural dynamo, and with it a magnetic field similar in shape to the field of a bar magnet (see also Chapter 11).

Many geologists now believe that the boundary between the core and the mantle is bumpy, with undulations of a kilometre or more extending over a few hundred to a few thousand kilometres. Some of them think that such undulations could help to explain not only variations in the daylength but also minor variations in the strength and direction of the planet's magnetic field.

Once geologists could measure very low magnetic fields they had a valuable surveying tool. Many rocks contain small quantities of iron. When these rocks were formed, the particles of iron acted as minute compasses and settled in the direction of the Earth's then magnetic field. The orientations of these minerals (**palaeomagnetism** or 'fossil magnetism') provide clues as to the reversals of the magnetic field. Fossil magnetism, however, holds another clue for geologists. The magnetic field is stronger at the poles than at the equator, and this affects the exact angle at which particles of iron are magnetized. By measuring the direction and dip of the magnetic field in a rock, geologists can determine the latitude at which the rock originally formed. Compare this with the present latitude of the rock, and you can build up a record of how land masses have rotated and moved relative to one another.

Oceans between continents
Sea-floor spreading

At the beginning of the seventeenth century the English philosopher Francis Bacon noted that the outline of the eastern side of the Americas and the western side of Africa looked as though they would fit together, rather like two huge pieces of jigsaw. In the centuries that followed, settlers in the New

World found huge deposits of coal in the American continent whose position seemed to match deposits on the European side of the jigsaw. Furthermore, scientists often found the fossilized remains of identical species of plants and animals on both sides of the Atlantic. Gradually, evidence trickled in that seemed to suggest that the land surface of the Earth was once a single huge continent, which broke up and slowly drifted apart to form separate units. The problem was that no one could explain how this happened.

Then in 1928 Arthur Holmes, a professor of geology at Durham University, suggested that there might be convection currents in the upper mantle. An American, Harry Hess, later extended this theory into the concept of **sea-floor spreading**. The idea was that convection would force molten rock, known as **magma**, to well up in the interior and to crack the crust above. As the magma cooled, it formed a strip of basaltic rock which would gradually spread out as more magma flowed from the crack. The great ridges that run through all the world's oceans were the sites of this activity.

Many scientists, and especially geologists, would not accept the idea of sea-floor spreading, until they were faced with data from magnetic surveys in the 1960s. Scientists on research ships had measured the magnetism of rocks across ocean ridges, such as the Mid-Atlantic Ridge. They found that the rocks on the ocean floor were magnetized in alternate directions in a series of bands parallel to the ridges. Moreover, the pattern of bands was identical on both sides of the ridge. They explained this by saying that when basaltic magma from the mantle cools on the ocean floor, it is magnetized in the direction of the magnetic field at that time. As the magma continues to erupt, it cracks the previous strip of newly solidified basalt, splitting it into two. If the magnetic field reverses, then this next strip of basalt will be magnetized in the opposite direction to the previous strip, and will form a band in between.

This, and the increasing age of the rocks as you move away from the ridges, supported the concept of sea-floor spreading. It showed how, as the mid-ocean ridges continually added

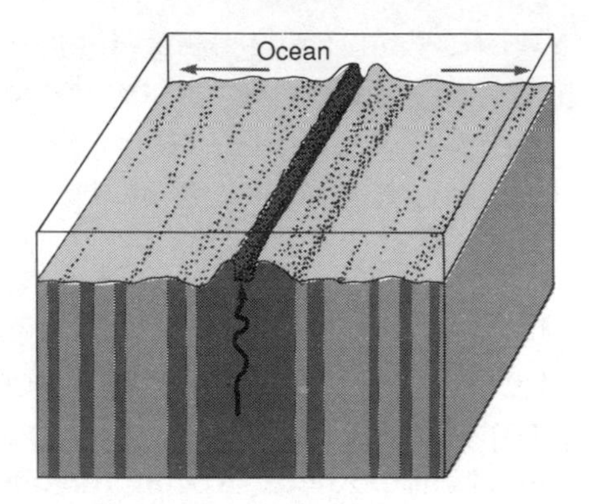

The rocks on either side of a mid-ocean ridge (right) are magnetized in alternate directions. This finding led to the theory that the lithosphere is made up of a series of plates (below)

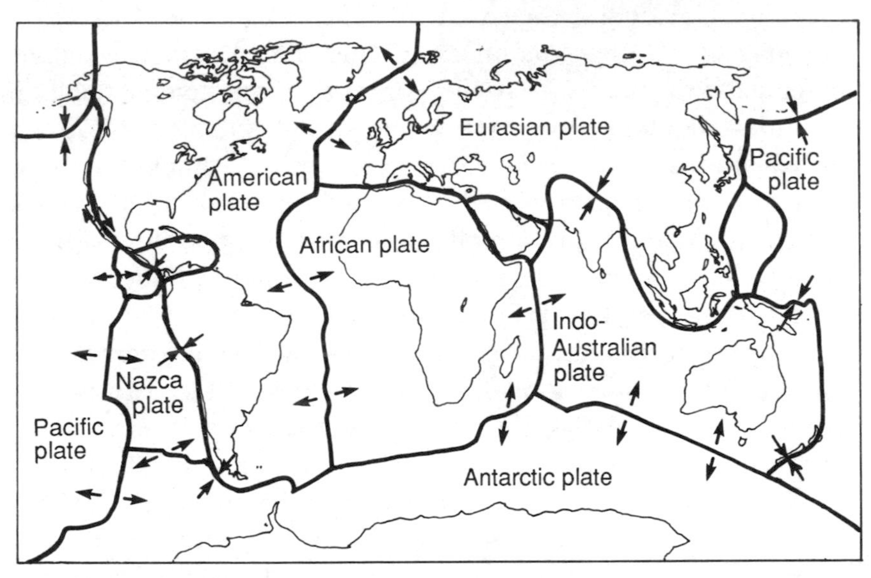

material to the ocean floors, continents once joined could become separated by huge oceans.

Plate tectonics
Global conveyor belt

In 1965 a Canadian geophysicist, J. Tuzo Wilson, brought together the ideas of continental drift and sea-floor spreading into a single global concept of mobile belts and rigid plates. Early in 1967 American geophysicists added another concept –

that of **underthrusting**, where one block of crustal material dips beneath another at deep-sea trenches. From such ideas, British, French and American scientists conceived a grand theory of the Earth's crust, and it soon took on the name of **plate tectonics** (the word 'tectonics' comes from the Greek word meaning 'builder'). The idea is that the outer layer of the Earth, the lithosphere, consists of six or more major plates which move over the hot, partially molten asthenosphere. As they do so, they bump into each other, move apart and slide past each other, carrying the oceans and continents with them.

The dynamics of plate tectonics

The theory of plate tectonics assumes that each plate behaves as a rigid unit, deforming only at its edges. The edges, or margins, can diverge, converge or slide past one another. But very little changes in the middle of a plate. In fact, some of the Earth's most dynamic features, such as volcanoes and earthquakes, define the boundaries of the main plates.

At mid-ocean ridges the plate

New plate forms at oceanic ridges, and the plates diverge. Where two plates converge, one slides steeply under the other, returning old plate to the asthenosphere

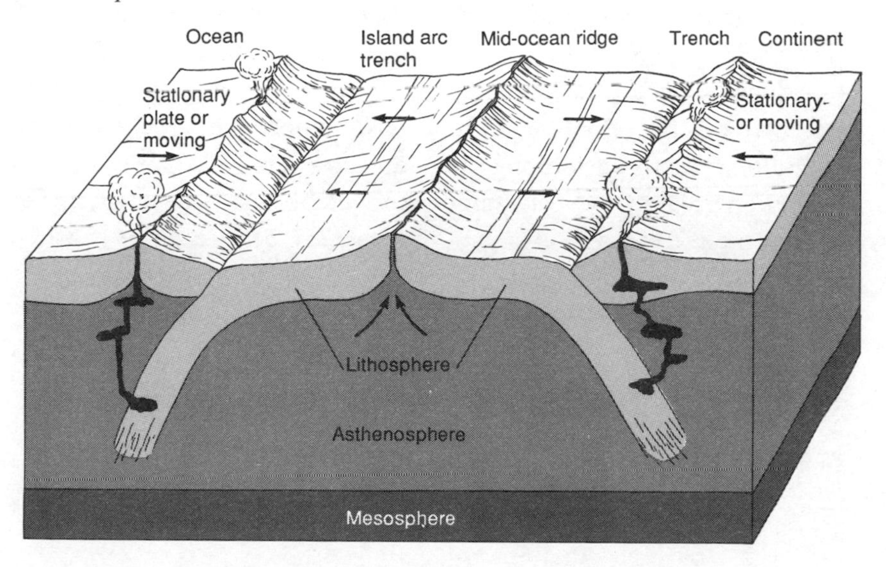

edges *diverge*, as magma, erupting from the asthenosphere, adds new plate. The ridges are like elongated, continuous volcanoes. Throughout their length shallow earthquakes (less than 50 kilometres below the surface) are common.

In zones where the plates run past each other the boundaries are said to be *sliding*. No plate material is added or destroyed at these **transform faults**, but they are associated with shallow earthquakes, sometimes of high magnitude. The San Andreas Fault of California is a classic example of sliding plates.

Where two plates *converge*, that is to say collide, a deep trench forms and one plate is deflected downwards into the asthenosphere. At these boundaries, mountains form and chains of islands appear. Earthquakes occur at a range of depths up to 700 kilometres (in fact, along the descending plate). Typical examples of the activity at converging plates are the Andes, and the Aleutian Islands which stretch out from Alaska.

Great mountain ranges are often the outcome of an ocean plate converging with a thick continental plate. The edge of the continent acts rather like the cutting edge of a plane – shaving great masses of material from the top of the ocean plate as it descends into the asthenosphere. Slowly, the 'shavings' build up to form mountains. Much of the descending ocean plate is reabsorbed into the asthenosphere. Some of the lighter material, during melting, rises up and swells the growing mountains by creating a 'root' beneath them. If this magma forces its way upwards, the result is a range of volcanoes.

Where a comparatively thin continental plate collides with an ocean plate, a deep trench develops as before. But the partial melting of the descending ocean plate causes volcanoes in the form of arc-shaped chains of islands.

Finally, two continental plates may collide. The rocks on the land are relatively light and too buoyant to descend into the asthenosphere. The result is a huge zone of crushing, with rocks and other materials being folded, overthrust and welded together. This is the process by which massive mountain chains such as the Himalayas emerge.

The key to plate tectonics is that the surface of the Earth exists in a state of equilibrium. Plates form at mid-ocean ridges where heat and material rise from the asthenosphere to form new crust. This is balanced by underthrusting, or **subduction** as it is now called, destroying plates elsewhere, returning older crust to the asthenosphere. In simple terms, new plate is driven to the surface by upward convection and old plate returned to the melting-pot, as it were, by downward convection.

Data gained by scientists using laser and satellite techniques now show that the various plates are moving between 1.5 and 7 centimetres per year (which is about the same speed as your nails grow). The plates that meet in the Mid-Atlantic Ridge, for example, are moving apart by 2 centimetres a year.

Plate tectonics provides a global framework that successfully explains many of the structural and geophysical phenomena on the Earth's surface, ranging from mountain-building and earthquakes to continental drift. But our knowledge of this layer is out of all proportion to the size of the planet.

25 February 1988

Further Reading

The Making of the Earth edited by Richard Fifield (Basil Blackwell, 1985) is a guide to the past 30 years of geological research, compiled from the pages of *New Scientist*. *The Illustrated Reference Book of the Earth* (Windward, 1982) edited by James Mitchell is a source of much inspiration for anyone wishing to know how our planet works. A thoroughly useful general reference book on geology is *The Cambridge Encyclopedia of Earth Sciences* (Cambridge University Press, 1982). Edited by David Smith, it includes contributions from many leading geologists. An invaluable specialized text at senior and university standard is Martin Bott's *The Interior of the Earth* (Elsevier, 1982). If you can visit the Geological Museum in London, 'The Story of the Earth' exhibition describes the Earth's crust and explains the key theories for understanding the structure of the Earth. An easy-to-read and illustrated handbook, *The Story of the Earth* (HMSO), complements the exhibition.

CHAPTER 11

The Earth's Magnetic Field

Nina Morgan

Our planet's magnetic field is in a state of constant change, and woe betide any navigator who is ignorant of this. Major upsets in the **geomagnetic field** can cause travellers to be misled by their compasses, and aeroplanes to end up flying far off course. Such **magnetic storms** can also wreak havoc in electrical power lines and radio communications. Because of the geomagnetic field, the north-seeking pole on a compass needle points to the north, and the south-seeking pole to the south. For magnets, north attracts south, and south north.

We know from records that during the past 160 years the strength of the Earth's magnetic field has fallen by 7 per cent. If it were to carry on in this way, it would disappear within the next 2000 years. But something like this has happened before. Geophysicists say that the present decline may be the start of a cycle that they have detected throughout the Earth's history and which is recorded in the rocks: a reversal of the Earth's magnetic field, in which north becomes south.

Magnetism takes its name from the mineral **magnetite**, an iron oxide which is naturally magnetic and which the Greeks mined at Magnesia (now in Turkey). The earliest written records we have of magnetism are from about 600 BC. The Chinese realized that small pieces of magnetite always pointed in the same direction if allowed to move freely. By placing a shard of the mineral on loosely packed straw, they made the first compass. The instrument evolved into a pivoted needle set in a box, known from Chinese records at least as early as the first century AD. By the twelfth and thirteenth centuries Western seafarers were using magnetic compasses. But it was not until

1600 that the British physicist William Gilbert realized that the Earth itself behaves as if it were a magnet.

The Earth has a magnetic field similar to that of a simple bar magnet. It is called a **dipole field**, because it has two magnetic poles at opposite ends of a line. We can visualize the field using lines of force which issue from the poles (see Chapter 2). The intensity, or strength, of the Earth's magnetic field is measured in terms of the force needed to hold a compass needle at right angles to its preferred direction, that is, pointing east-west, rather than north-south. Even at its strongest the Earth's magnetic field is only about 0.00005 tesla (the **tesla** is the SI unit of magnetic flux density). This is several hundred times weaker than the field between the poles of a toy horseshoe magnet. **Magnetometers**, used to measure the Earth's field, are so sensitive that a wristwatch upsets them.

Because the Earth acts in many ways like a bar magnet, it is tempting to imagine that it has a giant magnet buried at its core. This is not possible: materials lose magnetism as they become warmer, and it is too hot at the Earth's core for any material to retain its magnetism.

Magnetic types

Electricity and magnetism are linked. Both are associated with moving electrical charges, or currents. Electric currents produce magnetic effects and moving magnets induce electric currents. Magnetism is a directional force. We think of poles on opposite ends of a line, as on a compass needle. Like poles of a magnet repel and unlike ones attract each other. The familiar bar magnet sprinkled with iron filings demonstrates this very well. The force of attraction between two poles is proportional to the strength of the field and inversely proportional to the square of the distance between them.

The atoms of magnetic materials behave like tiny bar magnets. The magnetic properties of a material are determined by the behaviour of electrons around the nuclei of its atoms. An electron with spin moving around the nucleus of an atom produces its own magnetic dipole. Paired electrons with opposing spin cancel out each other's magnetic effects. Strong magnetic properties arise only when an atom has an electron in one of its inner orbits that is not paired up with one that has an opposite spin.

There are three main classes of magnetic material. The **ferromagnetic** materials, such as iron and nickel, are the best known. They will become magnetized when placed in a magnetic field, and most of them retain their magnetization away from the field. **Paramagnetic** materials, such as copper and oxygen, are magnetized only when they are in a magnetic field. **Diamagnetic** materials, which include superconductors, become magnetized in an opposite direction to a magnetic field imposed on them.

In any magnetized material, atoms are organized into tiny areas, or **domains**, in which the magnetic fields of the atoms are aligned. Normally, the directions of the magnetization of the domains are random. It takes energy to shift them so that they are aligned. If you tap an iron bar while it is in a magnetic field, it will become more magnetized. A material that has some magnetic domains aligned makes a better magnet than one in which the domains are randomly orientated. Materials lose their magnetism when heated because the extra jostling of the atoms knocks the domains out of alignment.

Ferromagnetic substances have a characteristic temperature, called the **Curie point**, above which they lose or change their magnetism. This is how scientists know that the nickel–iron core of the Earth cannot itself be a magnet. It is far hotter than the Curie point for a nickel–iron alloy.

A giant dynamo
Electric currents within

If there is no giant magnet within the Earth, what causes the magnetic field? Scientists believe that the Earth acts as a giant dynamo – the **geodynamo** – converting mechanical energy into electrical energy, and that the Earth's magnetic field is generated by electric currents deep within the planet (see Chapter 10). Our magnetic field probably comes from the slowly circulating mass of molten nickel–iron alloy which geologists believe makes up the Earth's 2200 kilometres thick outer core – a fluid layer sandwiched between the solid inner core and the mantle, which lies below the Earth's crust.

The liquid outer core flows in convection currents, driven by differences in density and temperature between the lower mantle and the solid inner core. Motion in this part of the core is affected by the Earth's rotation and the tidal pull of the Moon

Our magnetic field comes from nickel and iron in the outer core. At the surface, lava forms at ridges in the middle of the oceans. Black and white stripes show rock erupted in times of 'normal' and 'reversed' magnetization; the pattern is almost symmetrical

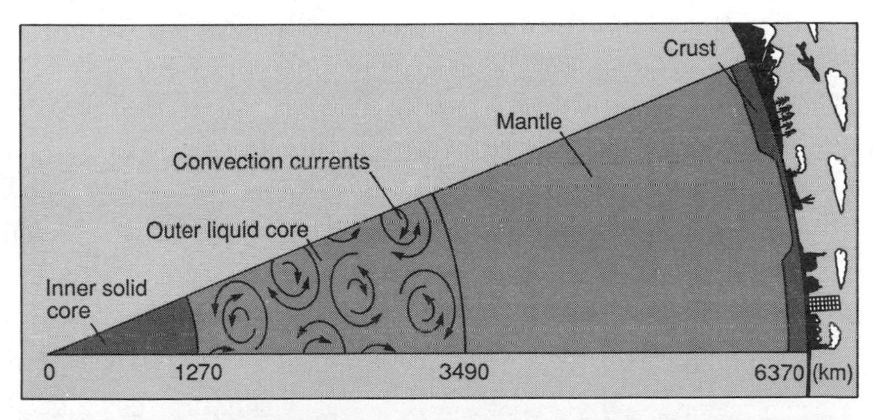

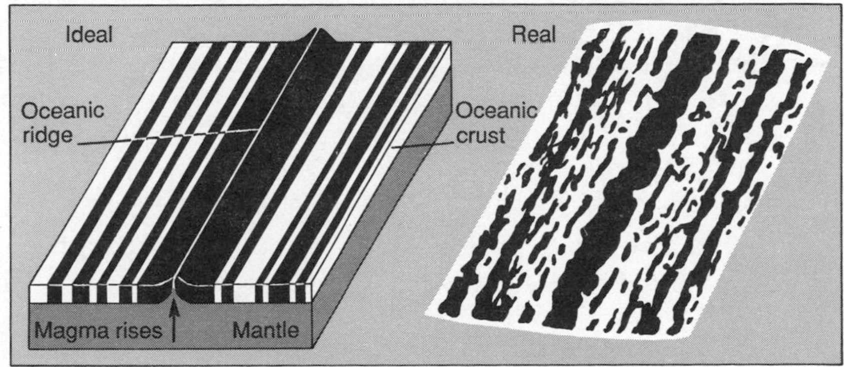

and Sun. Although the molten material is too hot to be intrinsically magnetic, it is a near-perfect electrical conductor.

Scientists think that the Earth's magnetic field arose from the Sun's field, when the planet formed from a cloud of dust swirling around the Sun. As conducting material passed through lines of force of the solar magnetic field, electrons experienced a force, and started to move, generating an electric current. The current, in turn, created a magnetic field, which was probably the start of the field we know now. It may be the flow of conducting material in the outer core that generates and maintains the geomagnetic field, and which now shields the planet from the solar field.

Scientists are confident that they understand the basic mechanism of the geodynamo. To get any further, they need more information about the Earth's interior. Sadly, they cannot, as Jules Verne might have had them do, undertake a journey to the centre of the Earth. Researchers try to understand the geodynamo by observing how the magnetic field changes at the surface.

Intensity, inclination and declination all vary, both on a daily basis and over periods covering tens, hundreds or even millions of years. These longer-term changes, or **secular changes**, are due to the convective movements which power the geodynamo.

Changes in declination can be dramatic, even over relatively short periods. For example, measurements made in London over the past 400 years show that the declination has varied from east to west and back again, a total change of more than 40 degrees. Navigators of all kinds, from hill-walkers to airline pilots, need to know about declination. Good maps show the direction of magnetic north relative to geographical north as part of the key. If you do not set the magnetic declination on your compass correctly when you set out, you could easily find yourself in the embarrassing position of the NATO paratroopers who mistakenly captured the wrong village during an exercise.

Shorter-term and much less predictable variations in the

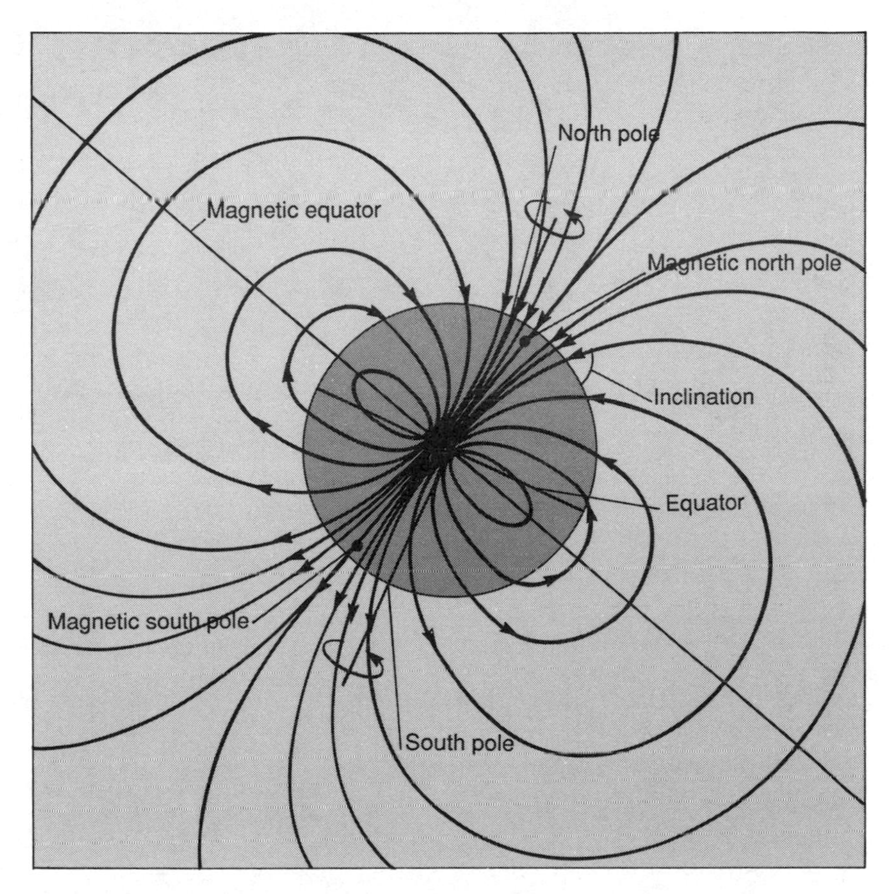

The Earth's magnetic axis is at present orientated at an angle of about 11° to the planet's axis of rotation. So the north and south magnetic poles, and the corresponding magnetic equator, differ from their geographical counterparts. The magnetic inclination, or dip, the angle between the lines of force and the surface of the Earth, varies at different latitudes. At the magnetic equator, the inclination is 0° and at the magnetic poles it is 90°. Magnetic declination, the angle between the direction of magnetic north (or south) and the corresponding geographical pole, also varies depending on where you are. This is because the magnetic and geographical axes of the Earth do not coincide, and because the magnetic poles constantly move about

Earth's magnetic field are caused by violent fluctuations, magnetic storms, in space. The Earth's magnetic field extends some 60 000 kilometres out into space. But the magnetic fields of some of the planets, such as Jupiter, are many times stronger and more extensive than that of the Earth, and can affect

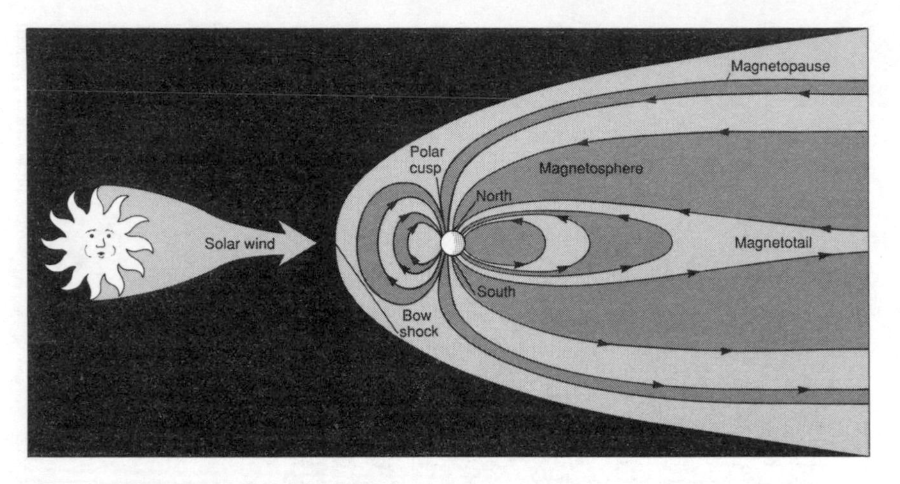

The Earth sweeps aside most of the charged particles and radiation from the Sun – the solar wind – because of the geomagnetic field. Some of the particles eventually reach the atmosphere by way of the polar cusps

it. The Sun, our nearest star, has the greatest influence over the Earth's magnetic field. The **magnetosphere**, the region dominated by the Earth's magnetic field, protects us from harmful solar radiation by deflecting the flow of **solar wind**, the stream of electrons and protons which flows away from the Sun at supersonic speeds.

The Sun rotates about once every 27 days. Its magnetic field also spins, probably slightly more quickly. Sunspots and solar flares emit extra radiation and charged particles which spiral away from the Sun, like water from a whirling hose pipe. When these extra bursts in the solar wind reach the Earth, they disturb the magnetosphere.

The consequences of disturbances during a magnetic storm can be spectacular, such as an aurora in the sky at high latitudes. All large disturbances upset telephone and radio communications, power supply lines, television transmissions, radar and radio navigation systems. In March 1989 a severe magnetic storm cut off electricity to six million people in Quebec, and put a nuclear power station out of action for 42 hours.

Finding the way with a compass

Orienteering by magnetic compass is the easiest method of navigation yet devised. You can use a compass at night or even in fog. Magnetic charts are the standard on which radio and satellite navigational aids are based. No matter how accurately the starting and finishing locations are known, it is essential to use a magnetic compass to set a course from one point to the other in all conditions.

Even modern, advanced navigation systems depend on the Earth's magnetic field to some extent. For example, gyrocompasses have a magnetic compass in their control systems to stabilize them and to counter the forces caused by the rotation of the Earth. Sophisticated navigational beacons, which work by comparing the phases of radio signals from two or more fixed transmitters, can be inaccurate when magnetic storms affect the ionosphere and shift the phase of radio signals. This is why all ships in excess of 150 tonnes must carry magnetic compasses and charts. Air traffic regulations require pilots to be able to navigate by magnetic compass to an accuracy of 0.5°. After all, an error of 1° could mean that an aeroplane would be flying 300 metres off course after only one minute.

In its simplest form a magnetic compass consists of a magnetized needle mounted so that it can rotate freely through 360°. The north-seeking end of the needle is free to point towards magnetic north. The great value of a magnetic compass is its inherent reliability, freedom from most kinds of interference, and simplicity.

Although the compass needle will always reliably point to magnetic north, the position of magnetic north is always moving as a result of fluctuations in the Earth's magnetic field. If you want to end up at a specific geographical location, you must know the local value of the magnetic declination. This is why an accurate map of the geomagnetic field is essential.

Modern geomagnetic maps are based on data from a worldwide network of observatories which measure the absolute value of the geomagnetic field every few days and continuously record its variation. Global charts of the geomagnetic field are produced every five years, but as the magnetic field is continuously shifting, the main

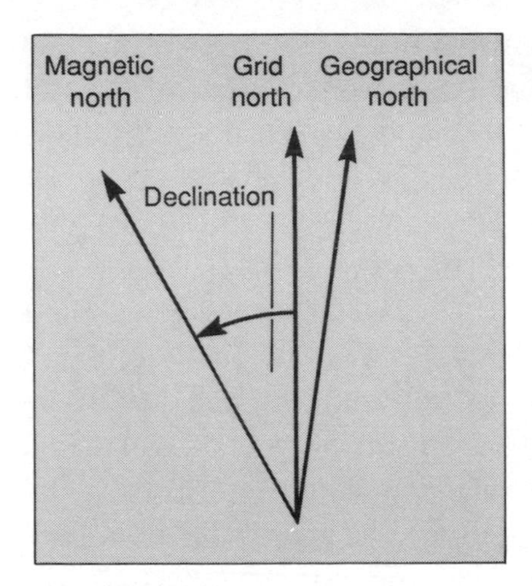

Always check the declination on your maps

field charts are out of date before they are published. Models of secular variation, the long-term variations caused by changes in convection currents in the Earth, are used to bring the charts up to date.

For normal navigating, no one uses the actual map of the geomagnetic field. The local information is extracted and included as a compass correction on topographical and nautical charts. But the in-flight computers of long-range aircraft can use the mathematical model directly. This navigation system is used on the American space shuttle.

Animals also navigate by the Earth's magnetic field. Experiments suggest that pigeons use the magnetic field to find their way home, and other birds use magnetism to tell them which way to migrate when they have no other guide, such as when the Sun is hidden by clouds.

Some animals, including bees, pigeons and salmon, have grains of magnetite within their bodies. At first, scientists thought that these grains acted like miniature compasses. But now it seems more likely that they allow the animals to recognize the pattern of magnetic changes, or anomalies, on the Earth's surface. The animals use these anomalies as a kind of 'road map'.

Biologists have found that pied flycatchers are sensitive to abrupt local changes in the Earth's magnetic field. Experiments with bees and pigeons have shown that when their in-built magnetic sensor is disturbed, for example by a magnetic storm, they actually become lost.

Although scientists have not yet fully established how this magnetic 'sense' works in animals, all the evidence points to its existence. It might even turn out that each location on Earth has a unique 'magnetic fingerprint'.

Magnetic disturbances can also cause compasses to deviate by several degrees, sending ships and aeroplanes far off course. On a spacecraft, exposed in space, the results are often devastating: magnetic storms can black out communications, burn out electronics and cause false triggers in command circuitry as well as expose astronauts to high levels of radiation. Magnetic storms can also make the Earth's atmosphere more dense, slowing satellites and shifting their orbits. This can seriously affect the accuracy of any navigation systems that depend on satellite passes. The satellites in the navigation system called **Navstar**, which commercial airlines use when landing, slow down and lose their precise positioning during intense storms.

Navstar satellites orbit so that three are always 'visible'

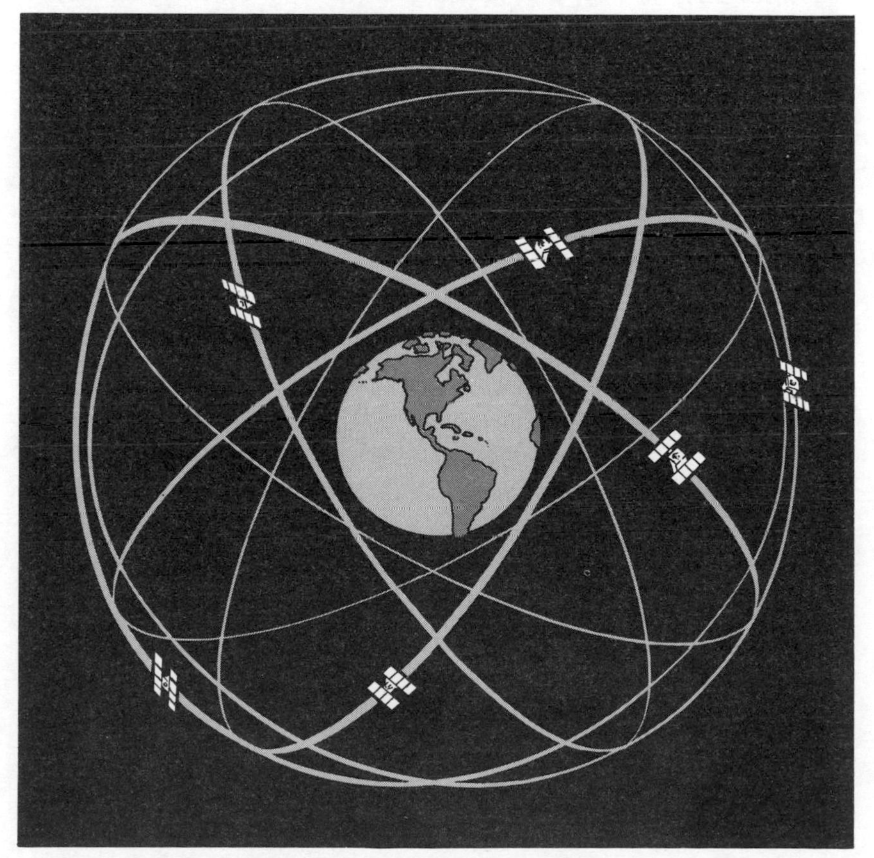

Although we cannot do anything to alter these potentially dangerous variations in the Earth's magnetic field, to be forewarned is to be forearmed. This is why in Britain, for example, the British Geological Survey provides scientists and industry with up-to-the-minute information on the constantly changing magnetic field, as well as daily forecasts of the level of geomagnetic activity.

Palaeomagnetism
Clues in the rocks

Although short-term changes in the Earth's magnetic field can be dramatic, to gain a longer perspective it is necessary to step back and consider the **palaeomagnetic**, or past magnetic, field that is preserved in ancient rocks. Minerals that contain iron become magnetized in the direction of the existing magnetic field as they form. For example, magnetite, the most magnetic iron oxide, crystallizes within igneous rocks as they cool. Rocks which cool and solidify quickly (close to, or at, the surface of the Earth), such as lavas, provide the most accurate 'snapshots' of the magnetic field in times past. Magma, or molten rock, cools more slowly when it solidifies far below ground, squeezed between other rocks. Such deep **intrusions** record the direction of the average magnetic field as the magma solidifies. Sedimentary rocks, such as sandstones, can also provide palaeomagnetic evidence. If magnetic particles are deposited in a sediment, they tend to orient themselves with magnetic poles parallel to the field lines at the time.

Scientists investigating palaeomagnetism have found reversals in the polarity of the Earth's magnetic field, so that magnetic north becomes magnetic south and vice versa. Such **polarity changes** have occurred relatively frequently throughout geological time. There have been at least nine major polarity changes in the past 3.6 million years. The most recent one happened only 730 000 years ago. Reversals in polarity also occur on the

Sun. The 11-year cycle of solar activity (such as sunspots) is in fact a reversal cycle of the magnetic field of the Sun.

Although polarity changes on Earth happened many times in the geological past, they did not happen instantaneously. Geophysicists now estimate that a typical reversal involves a transition period of about 1000 years. During that time the Earth's magnetic field is unstable, but no one knows what shape the field is, or exactly how it changes. Measurements of the field from rocks formed during ancient reversals indicate that the field may be only 10 per cent as strong as it is when the polarity is stable. Why do these reversals occur? What is happening to the Earth's magnetic field now? In recent years geophysicists have analysed data about the magnetic field recorded between 300 and 2000 years ago. The data support an idea that the British astronomer Edmond Halley investigated in the late seventeenth century. It looks as though the Earth's magnetic field drifts westwards at a rate of about 1° every five years. Geophysicists now believe that this drift could be related to a westward flow of the fluid outer core relative to the mantle. The rate at which the magnetic field is shifting suggests that the core fluid is moving at about 90 metres a day. Measurements of intensity show that the Earth's magnetic field is weakening and that it has decreased by about 7 per cent since 1845.

In the geological past, reversals followed weakenings of the magnetic field. The record of variation over the centuries not only confirms that the Earth's magnetic field is now weakening, it also provides evidence of a mechanism that could change the polarity of the geodynamo: an upset in the flow in the outer core. Are we now facing such a change? If so, what will happen to our shield against the solar wind? One thing is certain from the fossil record, life on Earth has survived such changes in the past. What will happen to human life and our civilization at the next change is less certain.

23 September 1989

Further Reading

The World of Physics by John Avison (Thomas Nelson, 1989) is a textbook that explains more about magnetism and how we live with it. *Understanding Magnetism, Magnets, Electromagnets and Superconducting Magnets* by Robert Wood (TAB Books, 1988) is a lively and up-to-date account of how and why magnetism works. *Earth's Magnetic Field* by Ronald Merrill and Mike McElhinney (Academic Press, 1984) is useful at university level. If you want to know how not to get lost, try *Maps and Compasses: a User's Handbook* by Percy Blandford (TAB Books, 1984) which explains all about navigation and maps.

CHAPTER 12

Acid Rain

Fred Pearce

Acid rain is not a recent discovery. A hundred years ago, Britain's first air pollution inspector, Robert Angus Smith, coined the phrase to describe the polluted rain in his home town of Manchester. He realized that the air in the city was not only filthy but also acidic, and that it was attacking vegetation, stone and iron. The finding, and the phrase, were forgotten, however, until the 1960s. Then Scandinavian scientists began to link pollution blown across the sea from Britain and the European mainland with acidified lakes and streams and the disappearance of fish from those waters. At first, no one could

The chain from pollutants to acidified lakes, disappearing fish and dying trees

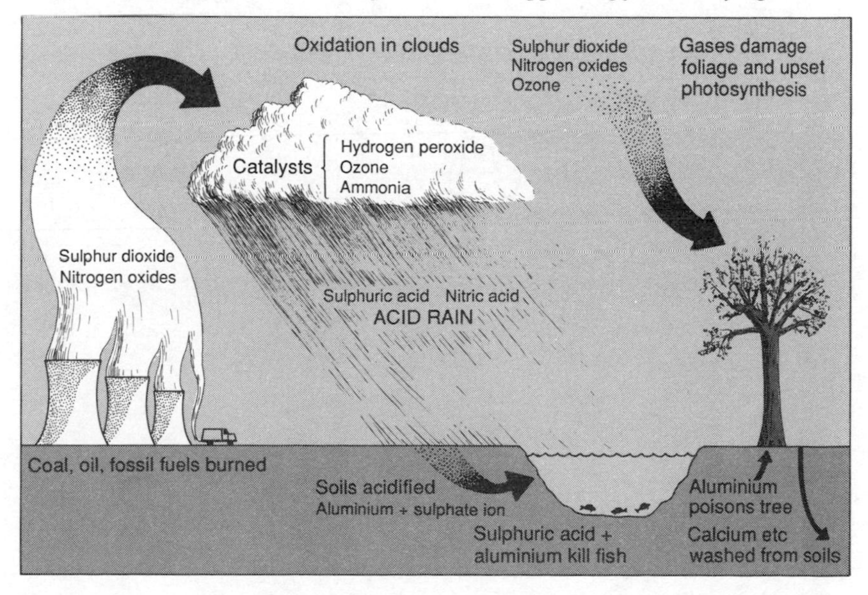

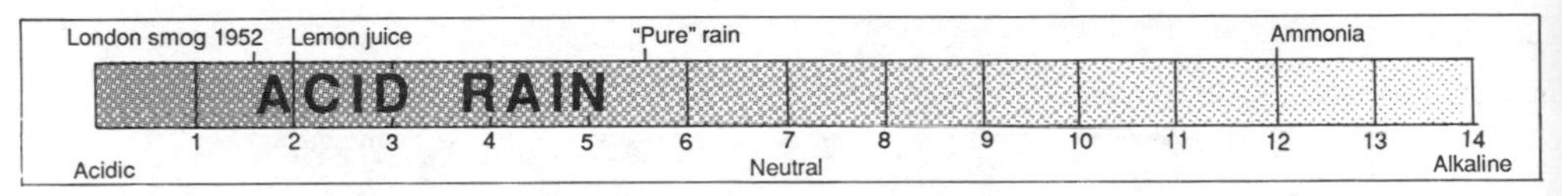

The pH scale measures acidity and its opposite, alkalinity. It runs from 1 (very acid) to 14 (very alkaline); 7 is neutral. The scale is logarithmic so, for example, rain with a pH of 4 is ten times as acid as rain with a pH of 5

explain why acid rain affected rivers when vast amounts of acids occurring naturally in soils did not. Still less had they established what killed the fish.

Rain is naturally slightly acidic – it reacts with carbon dioxide in the air to produce weak carbonic acid with a *p*H of about 5.6. In central Europe, rain is far more acid, with an average *p*H of around 4.1. Even on the western fringes of the

The total fallout of sulphur, in grams per square metre, over Europe in one year. Nitrogen shows a similar pattern

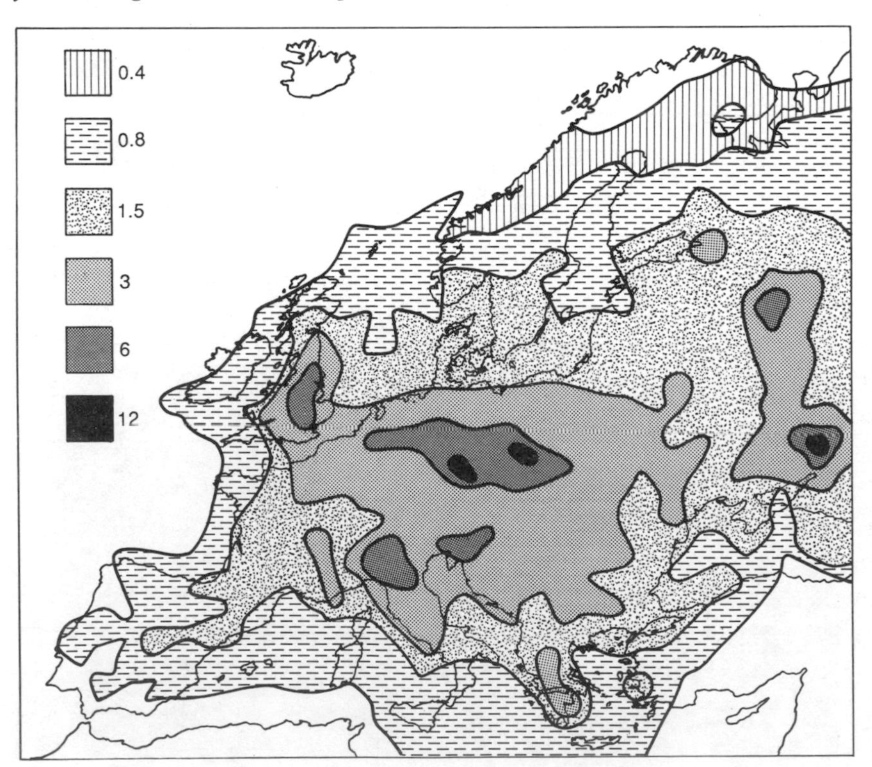

continent, such as Ireland and Portugal, pH averages 4.9. Rain from individual storms can have a pH of less than 3, and water droplets in fogs may be more acid still.

Acid rain has become an important political issue over the past 15 years, souring relations between the polluters and the polluted: between Britain and Norway, for instance, and the US and Canada. Air pollution can spread for thousands of kilometres across land and seas. Europe, the most concentrated seat of air pollution in the world, still shows the worst signs of acid damage, but there is evidence of damage from Australia to Mexico to China. Acid rain is now a global phenomenon. But until recently, the science of acid rain has been confused. Now, the key debate about soil chemistry and the transfer of acid from rain to rivers is largely over. Research into the atmospheric chemistry behind the formation of acid rain is beginning to yield important results. But the processes behind the destruction of Europe's forests – by far the most serious effect so far attributed to acid rain – are, to put it mildly, unclear.

In the clouds
A cocktail of chemicals

One of the chief culprits in acid rain is sulphur dioxide, which reaches the atmosphere by many routes. Sea spray, rotting vegetation, plankton and, in some places, volcanoes are important natural sources. Round the world, perhaps half the sulphur in the air comes from such sources and half from burning fossil fuels. Over Europe, the proportion from burning fuel is about 85 per cent.

Once sulphur dioxide (SO_2) reaches the atmosphere, it oxidizes in the air, reacting with the hydroxyl ion, which leads to the eventual formation of sulphuric acid (H_2SO_4). The process of oxidation can happen in dry air as a 'gas phase' reaction, but it is usually slow. A large mass of sulphur dioxide, in the plume from a power station chimney for instance, can travel hundreds of kilometres in stable air with little sign of conversion to acid. Over much of Europe most of the sulphur dioxide is absorbed by vegetation or building materials, unconverted. When sulphur dioxide is incorporated into clouds, however, oxidation by 'wet phase' reactions seems to happen much more quickly. Researchers tracking plumes from power stations across the cloud-covered Pennines found that most of the sulphur dioxide converted to acid within a couple of hours. Droplets in clouds are typically 10 times more acid than the rain that falls from those clouds.

There are grounds for believing that the whole atmosphere over Europe and the US is chemically much more reactive than a few decades ago, and that this accelerates the acidification. In heavily polluted urban air, tiny particles of metals such as iron and manganese catalyse the reaction. Elsewhere, gases such as ozone and hydrogen peroxide appear to be critical. One unexpected finding is that ammonia, given off in large quantities from open slurry tanks holding wastes from factory farms, may be an important catalyst, at least at a local level.

There are now strict controls on the emission of sulphur dioxide from power stations in the US and most European

Sources of sulphur dioxide emissions in Britain

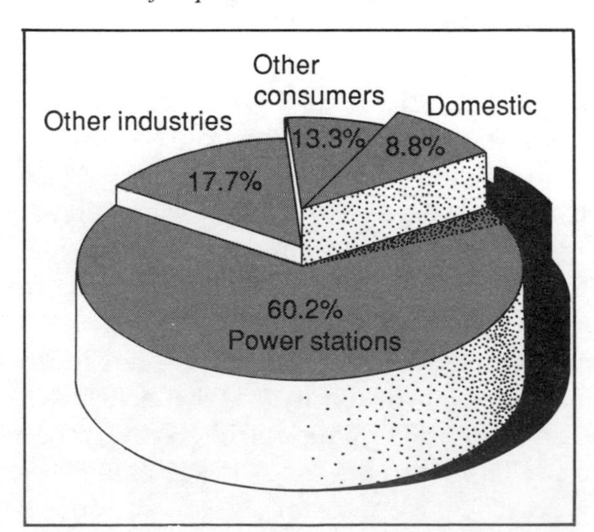

A history of assault in Britain

Britain has an instructive history of assault by acid rain. Ever since the chimney became commonplace, ejecting fumes from burning fuel into the outside air, towns have suffered from air pollution. Smoke has been the most obvious problem. But sulphur, too, has had a part to play. Coal contains, typically, between 1 and 3 per cent sulphur, which disappears up the chimney as sulphur dioxide.

In December 1952 a pollution disaster hit London. A deadly 'smog' – a cold, black and sulphurous stew of air – hung over the city for almost a week, trapped by a blanket of warm air. It was the worst of a long series of 'pea-souper' smogs to have hit London. It killed about 4000 people.

Smog irritated bronchial tubes, which became flooded with mucus. People choked to death or suffered heart attacks as they fought for breath.

At the time, doctors blamed the copious quantities of smoke and sulphur dioxide that accumulated in the smog. Scientists now believe that the formation of highly acid particles may have been important. One estimate puts the *p*H of the London smog of 1952 at 1.6, rather more acid than lemon juice.

After 1952 a public outcry led to legislation banning the burning of smoky fuels in most towns and cities. This, coupled with the arrival of cheap North Sea gas, the growing use of electricity and a decision by the government to build the next generation of power stations outside urban areas, ensured that towns became vastly cleaner places.

But the government took no specific steps to limit emissions of sulphur dioxide from power stations burning coal and oil. Britain's output soared. By the mid-1970s, chimneys up to 300 metres high were putting more than 5 million tonnes of sulphur dioxide a year into the air over Europe. Other European countries emitted similar amounts. The result was that towns were cleaner, but sulphur dioxide spread in increasing amounts to the remotest corners of the continent. As later emerged, rainfall everywhere became more and more acid.

countries. There, the rising stars of air pollution are nitrogen oxides, which are producing acid rain on a scale approaching that of sulphur dioxide. Nitrogen oxides are given off by power stations (again) and car exhausts, and include both nitrogen dioxide (NO_2) and nitric oxide (NO). In the air, nitric oxide quickly converts to nitrogen dioxide, which is itself oxidized and changed into nitric acid (HNO_3).

In the soil
Chemicals on the move

Many of the thin soils that cover granite or sandy rock in northern Europe, Canada and parts of the US have been acidic ever since they were formed 10 000 years ago at the end of the last ice age. Unable to neutralize acids that arise from natural processes, many of these soils contain the equivalent of thousands of years of deposition. When the Scandinavians first claimed that pollution from Britain was causing acid lakes, critics pointed to this. It was areas with these soils that had many of the most acid lakes and rivers, so, they asked, who needs acid to explain what is going on?

This theory weakened in the early 1980s when scientists reconstructed detailed histories of acid lakes. They looked in lake sediments for remains of tiny organisms called diatoms. These are very sensitive to acidity. It quickly emerged that lakes in Scandinavia, Scotland and Canada had become acid only after the height of the Industrial Revolution in the nineteenth century. The clincher was a detailed study of soil chemistry. Natural acid in soils is dominated by carbonic and organic acids. But to transfer acidity to surface waters that drain the soil requires mobile negative ions to bind to the acid hydrogen ion. The carbonic and organic acids do not contain such ions. The strong acids found in acid rain do, however. In particular, the negative sulphate ion in sulphuric acid is mobile within the soil and efficiently transfers acidity from soils to surface waters. The nitrate ion, too, would behave in this way if it were not normally taken up by plants first.

In the water
The poisoning of fish

Fish, notably brown trout and salmon, have disappeared from thousands of lakes and several large rivers in southern Scandinavia since the 1950s. In 1900 anglers caught 30 000 kilograms of salmon in the seven main rivers in southern Norway. Since 1970 no salmon have been taken. Many lochs in Scotland, especially in the Galloway Hills, are also fishless – as are hundreds more in Canada and parts of the eastern US. The fish died or failed to reproduce in acid waters, but only rarely is acidity the cause.

The deaths are usually due to poisoning by aluminium. All soils contain massive amounts of aluminium. Normally, though, it is in insoluble form, bound to the soil. But, just as the sulphate ion in acid rain transplants acid to rivers, so it can 'unhook' aluminium from its complex compounds and wash it into streams. There it interferes with the operation of the fish's gills, so that they clog with mucus. This reduces the amount of oxygen that reaches the blood. The mixture of aluminium and acid in lakes and rivers has a profound effect on freshwater

How the acidity of a loch in Scotland soared with industrialization

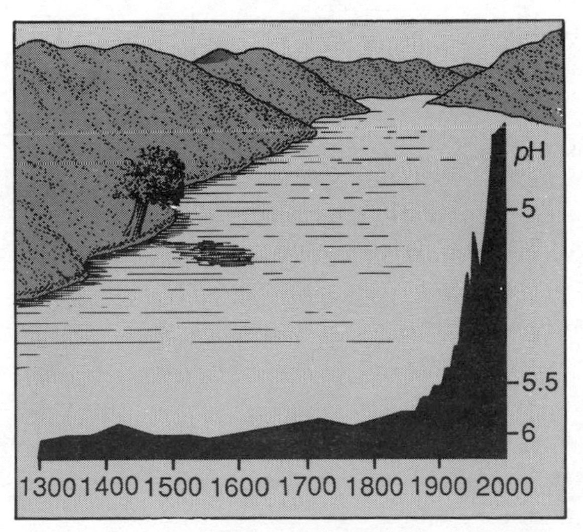

ecology. Acid lakes are usually crystal clear with luscious carpets of green algae and moss. All this green is deceptive. When algae and moss proliferate, they change the 'metabolism' of the lake. They slow down the decomposition of the greenery, providing less energy for the life there. The result is a changed ecosystem, with fewer species.

Canadian scientists tried dosing a lake deliberately with acid over several years. The lake initially had a *p*H of 6.5. At a *p*H of around 6, shrimps and minnows were the first organisms to disappear. Trout eat minnows so young trout soon failed to appear. At 5.6 the external skeletons of crayfish softened and were soon infested with parasites. Their eggs were overrun by fungi. Soon there were no crayfish.

In the forest
Trees under attack

The first signs of the decline of Europe's forests appeared in parts of the Alps, in the mid-1970s, when fir trees started to lose their needles. Then, in west Germany, the crowns of Norway spruce began to thin and needles turn brown. German foresters decided in 1981 that the time had come to make their warnings public in newspapers and magazines. After 1984 the decline stabilized in Germany. In other countries the forests continued to deteriorate, with deciduous trees increasingly affected. In 1986 a European study classified about 29 per cent of trees in the Netherlands as moderately or severely defoliated. West Germany had 20 per cent in this category, and Czechoslovakia and Switzerland 16 per cent. A survey of selected sites in Britain recorded 29 per cent. What is doing the damage?

In Germany, scientists pointed early on to acid soils. Soils in parts of Sweden, Germany and Britain are known to have become more acid in recent decades. Acid waters draining from the soils wash out nutrients and liberate aluminium, which the roots of trees may take up. Without essential nutrients, such as magnesium and calcium, trees starve to death.

Sulphur dioxide also directly damages leaves and needles – it blocks the stomata on leaves, preventing photosynthesis. Ozone derived from vehicle exhausts reaches levels each summer that are toxic to some trees, especially in conjunction with sulphur dioxide. Shoots seem to develop at the expense of roots, photosynthesis is disrupted, and chemical processes in general upset. The picture slowly emerging from a mass of frequently contradictory research is one in which acidifying soils and direct attack from air pollutants may make trees more vulnerable to assault.

Severe frosts may initiate decline. The suggestion, though, is that several air pollutants, including ozone and sulphur dioxide, reduce the frost-hardiness of plants. Air pollutants also encourage fungi and pests such as the bark beetle to grow. The ambrosia beetle is attracted by chemicals such as terpenes given off by trees under stress. Ammonia, as we have seen, efficiently oxidizes sulphur dioxide to create the sulphate ion. The resulting ammonium sulphate often forms on the surface of vegetation. Ammonia reduces frost-hardiness; ammonium sulphate, when it reaches the soil, creates both sulphuric and nitric acid.

Some researchers believe that doses of nitrogen, in the form of nitric acid, nitrogen oxides or ammonium compounds, reach the soils in some parts of Europe in such quantities that they 'saturate' the soils. We always assume that nitrogen does nothing but good to plants, but in excess it may stress trees by forcing them to grow when they are short of nutrients.

In the future
Prospects for a clean-up

By the beginning of the next century the output of sulphur dioxide and nitrogen oxides from industrialized nations will be much lower than today. Most countries will install chemical plants to remove the sulphur from emissions from new power stations before they reach the atmosphere. A typical process involves large quantities of limestone (calcium carbonate) to remove sulphur dioxide and form gypsum (calcium sulphate).

The bad side of ozone

Ozone is an enigma. In the upper atmosphere it is a 'good thing'. It shields the Earth from ultra-violet radiation. Close to the ground ozone is a hazard. It damages plants and many materials from rubber to textiles; it hastens the formation of acid rain; and it may trigger asthma attacks and bronchitis.

Ozone is formed in sunlight by nitrogen oxides and traces of hydrocarbons in the air. Motor vehicles produce both – over Britain, around two thirds of the ozone is generated by vehicle exhausts. Power stations are among the other sources of nitrogen oxides. Hydrocarbons come from everything from industrial solvents to the methane from ruminating cattle and leaking North Sea gas.

Not surprisingly, the amount of background ozone close to the ground has roughly doubled over Europe in the past three decades.

There is a debate about how best to reduce the formation of ozone. The amount of the fastest-reacting hydrocarbons, such as alkanes and alkenes emitted by vehicle exhausts, probably accounts for the peak concentrations of ozone in the summer. But there are also slow-acting hydrocarbons, such as methane, in the atmosphere which may take years to react with nitrogen oxide and create ozone. They are so common that ozone may best be stemmed by controlling nitrogen oxides.

Ozone is but one of the chemicals produced by reactions between pollutants in sunlight in industrial areas and which accelerate the formation of acid. One consequence of this increasingly reactive chemical soup in the atmosphere is the formation of heat haze, which is usually an aerosol, or 'mist', of sulphates and nitrates. Such a soup, near Los Angeles, recently produced a fog with a *p*H of 1.7.

Careful design of combustion processes inside plants will reduce the output of nitrogen oxides from power stations.

Catalytic converters, which contain platinum catalysts, are being bolted on to exhaust systems of vehicles to remove hydrocarbons and nitrogen oxides. They are fitted as standard on cars in many countries, notably the US and Japan, and reduce the formation of both acid rain and ozone. The use of cars, however, is rising so rapidly that even stringent controls may do little more than maintain the status quo. Even worse is that, if all air pollution were halted tomorrow, many of the effects outlined here would remain for decades. Some soils would retain their man-made acidity and reservoirs of sulphur. Acid and aluminium would continue to poison lakes and streams. A British attempt to model acidity in lochs in Scotland predicts that even a halving of acid fallout in the hills by the year 2000 would only maintain the acidity of water at current levels. The only way to ameliorate the effects of acidity in the short term is to combine a clean-up of the air with dumping vast quantities of limestone in soils and waters and adding fertilizers and nutrients to forest soils. One leading German scientist estimates that such a programme would cost his country £15 000 million. It is a pessimistic picture, but at least research is beginning to unravel the complicated chemistry of acid rain.

5 November 1987

Further Reading

Acid Rain by Fred Pearce (Penguin, 1987) is a popular account of the causes of acid rain, the damage it has done, and the remedies which need to be taken. *Ecological Effects of Deposited Sulphur and Nitrogen Compounds* (The Royal Society, 1984) is the collected papers from a meeting at the Royal Society, and provides one of the best overviews of scientific knowledge in this complex subject. The text should be accessible to those with the equivalent of A-level biology or chemistry.

CHAPTER 13

The Ozone Layer

John Gribbin

Earth is unique among the planets of our Solar System in having an atmosphere that is chemically active and rich in oxygen. Chemical reactions ought to reduce this atmosphere in a few thousand years to an unreactive state, with oxygen becoming locked up in stable compounds such as carbon dioxide and water. This does not happen because life on Earth is constantly renewing the oxygen content of the atmosphere. All the other planets that have atmospheres are surrounded by blankets of inert gases, such as carbon dioxide, hydrogen and methane. No interesting chemical activity can be going on there, because there are no active chemicals available for reactions. Alien explorers visiting our Solar System from another star would be able, by spectroscopic techniques, to pick out the peculiarity of the Earth and identify it as a likely home of life long before they passed the orbit of Pluto.

Life on Earth
The oxygen-giver

Oxygen exists in the atmosphere because there is life on Earth; and life exists on the land surface of the Earth because there is oxygen in the atmosphere.

When the Earth formed, as a molten ball of rock, it had no atmosphere. As the Earth began to cool, the atmosphere developed from gases emitted by the hot surface of the young planet, and from volcanoes. The gases emitted by volcanoes today are chiefly water vapour (80 per cent) and carbon dioxide

(12 per cent), plus sulphur dioxide (7 per cent), nitrogen (1 per cent) and traces of other substances. The mixture of gases emitted when the Earth was young was probably much the same as this, and we can assume that carbon dioxide dominated the first atmosphere of our planet.

At that time, 4.2 billion years ago, ultraviolet radiation in sunlight would have reached the surface of the Earth, keeping it sterile. Ultraviolet light, with wavelengths from below 200 nanometres up to 350 nanometres (a nanometre is one thousand millionth of a metre), is a powerful sterilizing agent and hospitals use it today, for example, to kill microorganisms on surgical equipment. However, ultraviolet radiation may have played a part in helping life to become established. A little ultraviolet light would provide energy to encourage the kinds of chemical reaction that scientists believe were the precursors to life. Some people suggest that conditions would have been just right at a depth of about 10 metres below the surface of primordial pools of water – just deep enough to filter out most of the ultraviolet light, but shallow enough to leave sufficient ultraviolet light to encourage chemical reactions.

Whatever triggered off life on Earth, oxygen was a poison to the early forms of life. Organisms used energy from sunlight to break down carbon dioxide so that they could use the carbon in building their own cells; however, they had to dispose of oxygen as a waste product. By about 2.5 billion years ago there were large quantities of oxygen in the atmosphere. Many organisms were poisoned by this 'toxic' gas; others evolved defences to shield themselves from its reactive properties. Many other forms of life found ways to make use of oxygen.

The presence of oxygen changed the course of evolution in two ways. First, because oxygen absorbs ultraviolet radiation, life could now colonize the land surface of the Earth, beneath the shielding blanket of an oxygen-rich atmosphere. Secondly, because oxygen is so reactive, life forms that became adapted to using oxygen, combining it with carbon from their food to release energy, developed more energetic lifestyles than the first organisms, which depended on sunlight for their energy. Simple animals appeared only after plants had converted a significant amount of the carbon dioxide in the atmosphere into oxygen.

The atmospheric blanket
Atoms in combination

Today, the atmosphere of the Earth is about 75 per cent nitrogen (by mass), 23 per cent oxygen, 0.04 per cent carbon dioxide and 1.3 per cent argon. The structure of the atmosphere is most simply described in terms of the way temperature varies, from the ground up. The surface of the Earth is warmed

The ozone layer forms part of the stratosphere. It exists because oxygen filtering up from the top of the troposphere reacts under the influence of sunlight to form ozone. This 'photodissociation' of oxygen is greatest above the equator and the tropics because that is where solar radiation is strongest and most direct. From these regions, ozone is transported by winds within the stratosphere around the Earth, towards the polar regions to maintain the ozone layer

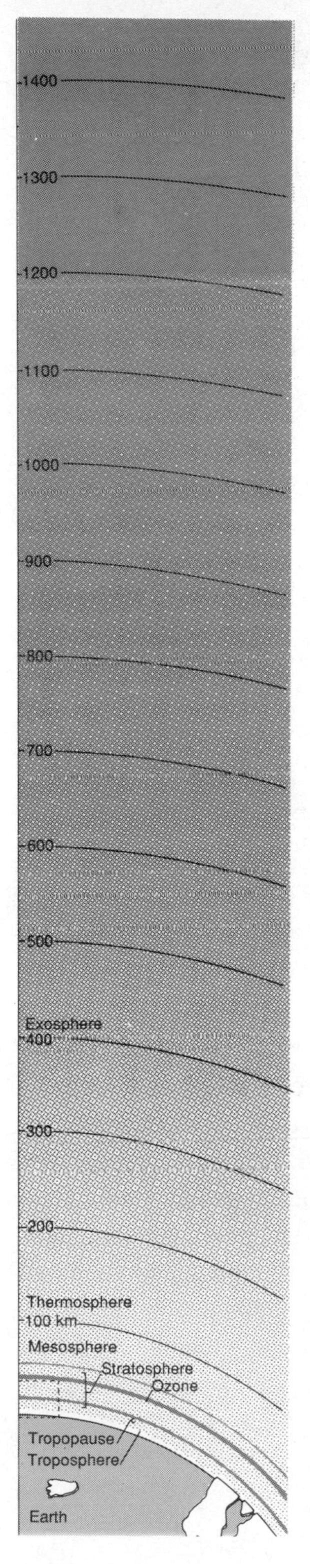

by sunlight, most of which passes unaffected through the atmosphere. This energy is mostly in the form of yellow light, with wavelengths between 500 and 600 nanometres. The warm surface of the Earth (land or sea, trees or car roofs) radiates heat back to the atmosphere at longer wavelengths, however, in the infrared part of the spectrum. This infrared heat is absorbed by water vapour, carbon dioxide and other molecules in the lower atmosphere, making it warm near the ground.

The warmth decreases with increasing height. At an altitude of about 11 kilometres the temperature drops to −60 °C; then it starts to become warm again (the cooling actually stops at about 8 kilometres over the poles and at around 16 kilometres over the equator). The air below the boundary between warming and cooling layers is the troposphere, the layer of the atmosphere in which weather occurs. But from this altitude up to about 50 kilometres, the atmosphere gets warmer with increasing height due to a heat source associated with the presence of ozone in the atmosphere. Increasing warmth means that energy is being absorbed; that energy is solar ultraviolet radiation, and it is being absorbed in the ozone layer of the atmosphere.

Profile of the Earth's atmosphere

Ozone is a form of oxygen which has three atoms in each molecule (trioxygen, O_3), compared with the usual form of oxygen, the kind we breathe, which has two atoms in each molecule of the gas (dioxygen, O_2). Ordinary oxygen molecules (whenever I use the term 'oxygen' I mean dioxygen) are broken apart by ultraviolet radiation wavelengths below 242 nanometres:

$$O_2 \xrightarrow{uv} O + O$$

The free oxygen atoms produced in this way are very reactive. Some will react with other oxygen molecules to make ozone. This can happen only if some other molecule is present to take up the kinetic energy released in the reaction; usually, this other molecule is nitrogen, but since it could be almost anything, it is convenient to label it M:

$$O + O_2 + M \rightarrow O_3 + M$$

As well as making ozone, this pair of reactions provides energy to molecule M, making it move faster. When molecules of gas move faster, the gas becomes hotter. So solar ultraviolet is absorbed, converting oxygen into ozone and warming the stratosphere.

Ozone itself reacts under the influence of ultraviolet light at longer wavelengths, especially between 230 and 290 nanometres:

$$O_3 \xrightarrow{uv} O + O_2$$

Most of the single oxygen atoms produced like this combine with other oxygen molecules to re-form ozone. The composition of gases in the atmosphere is not changed, but more ultraviolet has been absorbed, and the stratosphere has become warmer.

Other reactions do destroy ozone more permanently. For example, oxides of nitrogen act in a catalytic way to convert ozone and oxygen atoms into molecular oxygen:

$$NO + O_3 \rightarrow NO_2 + O_2$$
$$NO_2 + O \rightarrow NO + O_2$$

with the net effect:

$$O_3 + O \rightarrow O_2 + O_2$$

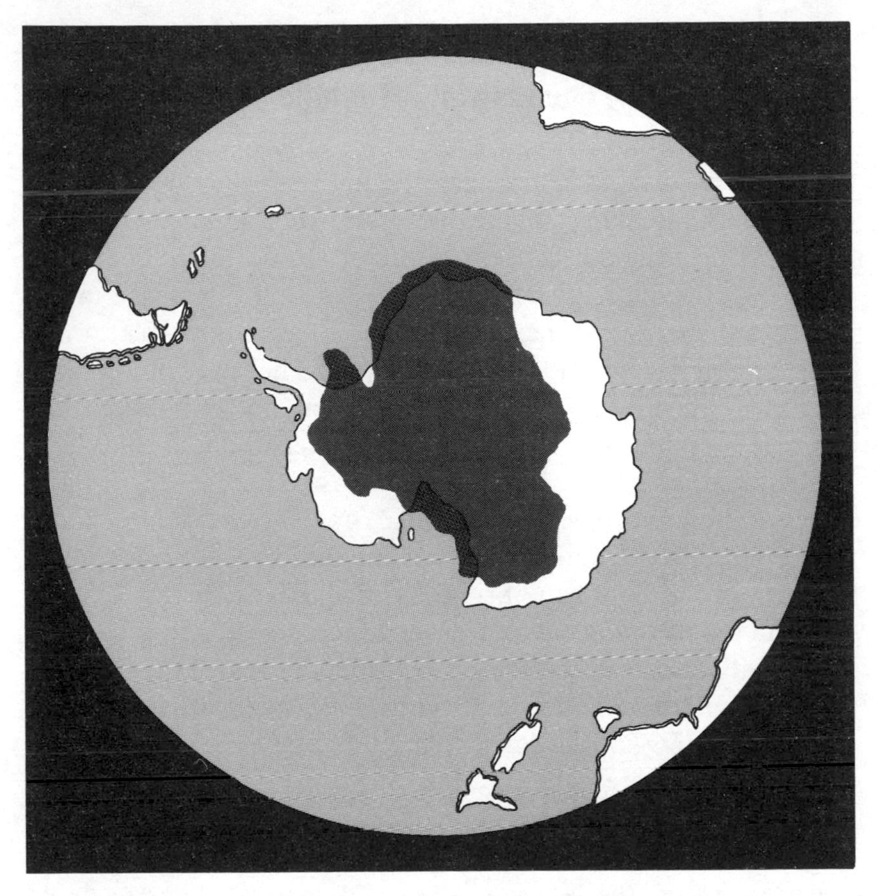

In October 1987, when the hole in the ozone layer (dark grey) above Antarctica was at its biggest, it covered an area the size of continental United States

Because NO is returned at the end of this cycle, it can 'scavenge' many molecules of ozone before eventually being converted through NO_2 to nitric acid.

The overall result of these and other reactions is that ozone is constantly being produced and constantly being destroyed in the stratosphere, in a dynamic balance. The images that people sometimes have of the ozone layer as being like a soap bubble that can be popped so it is gone for ever, or of a resource like oil that may run out one day, are incorrect.

Human activities
The making of a hole

Although ozone is constantly being manufactured in the stratosphere, human activities may still tilt the balance of the reactions so that the concentration of ozone falls. The usual analogy concerns a tub of water that is being filled by a tap at a constant speed, but the tub has a hole in its bottom through which water escapes at the same speed. The level of water in the tub (equivalent to the concentration of ozone) stays the same. If we make the hole in the tub bigger, the first thing that happens is that more water flows out. As it does so, the level of water in the tub drops. As the level drops, the pressure of water at the hole drops, and the flow from the tub slows down. At some point, a new equilibrium is reached, with the level of water in the tub constant at a lower level.

Present concern about the ozone layer centres on the possibility that human activities may be making the 'hole in the tub' bigger, speeding up the rate at which ozone is being destroyed in the stratosphere, while the production rate stays the same. This could happen if molecules such as NO are produced by human activities and reach the ozone layer. There, as the equations show, the NO molecules will increase the scavenging of ozone. The first concern about possible human impacts on the ozone layer came in the 1970s when plans were being made for large fleets of supersonic aircraft, flying higher than Concorde. The jet engines of such aircraft, sucking in air to provide the oxygen they needed, would inevitably have produced NO (and other oxides of nitrogen) since so much of the air is nitrogen. These aircraft never flew, but that concern might still apply if hypersonic vehicles, such as HOTOL, were ever to leave the drawing board. At present, however, concern focuses on chlorine. Chlorine atoms (Cl) do exactly the same scavenging job as NO:

$$Cl + O_3 \rightarrow ClO + O_2$$

$$ClO + O \rightarrow Cl + O_2$$

with the net effect:

$$O_3 + O \rightarrow O_2 + O_2$$

and Cl being returned to go round the cycle thousands of times.

The sources of the chlorine that cause concern are compounds known as chlorofluorocarbons, or CFCs. These are widely used in refrigeration (as the working fluid in the pipes of a fridge), air-cooling systems, to make the bubbles in foamed plastic, to a declining extent in aerosol sprays, and as cleansers in the computer industry. CFCs pose a potential

The drop in the mean total ozone over Halley Bay for Octobers from 1957 to 1985 matches closely the build-up of CFCs (here the major ones F11 and F12) in the southern hemisphere. (Note the reverse scale for the CFCs)

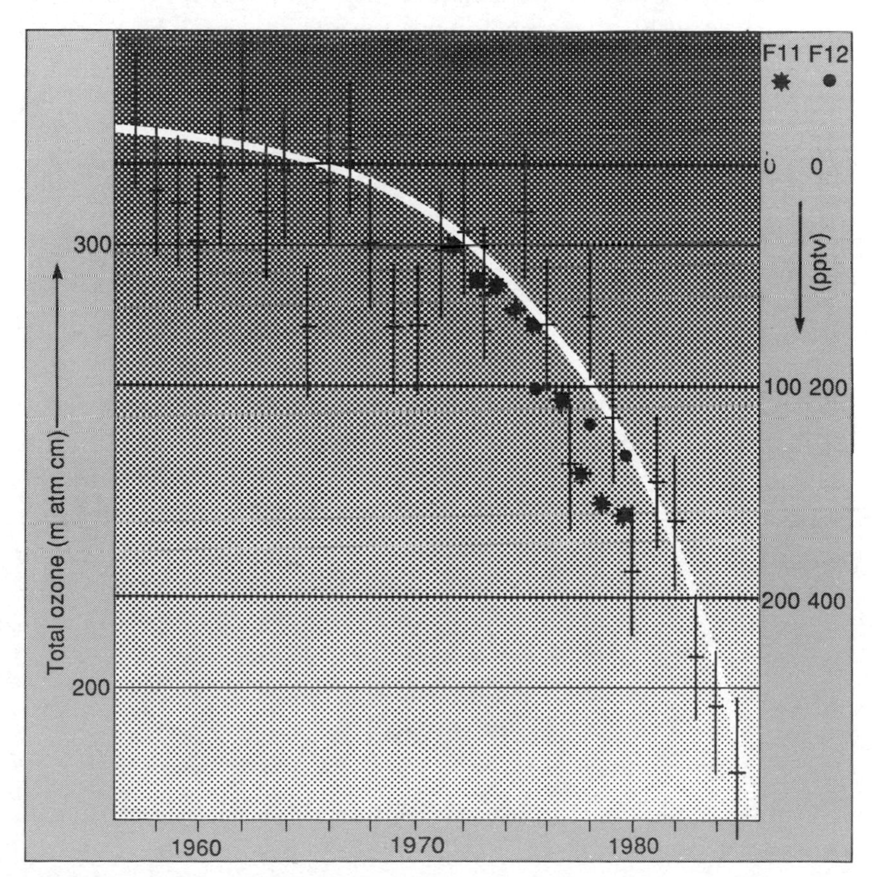

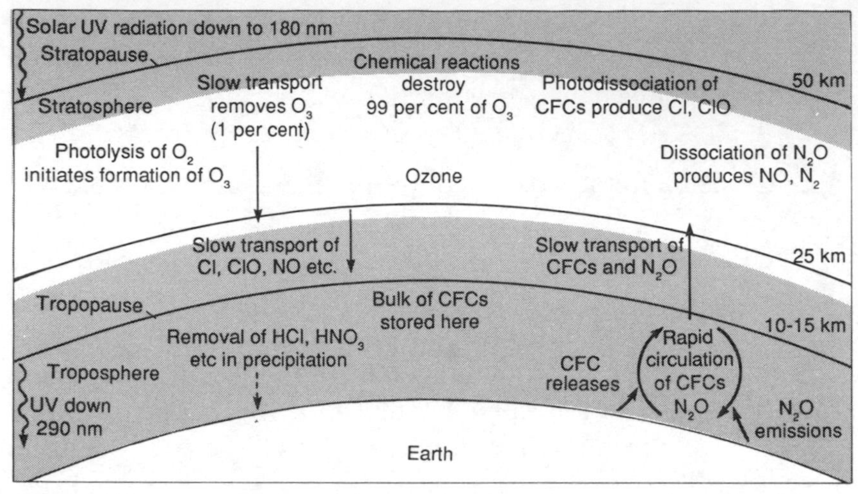

Numerous processes determine the concentration of ozone in the stratosphere

threat to the ozone layer because the ones widely used today are very long-lived (some longer than 100 years) and spread throughout the atmosphere. Some must inevitably reach the stratosphere, where they are broken down by solar ultraviolet radiation, releasing chlorine atoms.

Until recently, it seemed that this was of no immediate concern – a problem for the twenty-first century perhaps. Calculations of how ozone might be depleted as chlorine built up in the atmosphere suggested a slow, steady decline in ozone concentrations of a few per cent over several decades. But the discovery of a 'hole' in the ozone over Antarctica shows that the atmosphere may respond in a 'non-linear' way to this kind of disturbance. What seems to have happened over Antarctica is that the build-up of chlorine had very little effect on the ozone layer until some critical threshold was reached. Then, a very small increase in the amount of chlorine caused a very big change in the chemistry of the stratosphere.

Radiation
Damage to life?

The unique meteorology of the Antarctic means there is little prospect that the hole in the sky there will spread to cover the rest of the world. But there is now concern that some separate non-linear effects may be at work in the other parts of the world, and that other holes in the ozone layer might appear, most probably over the North Pole, but possibly anywhere. This is a cause for concern – because solar ultraviolet radiation would reach the ground in increased quantities if the ozone layer were depleted. Some ultraviolet does reach the ground today. It is in a waveband known as UV-B, from 290 to 320 nanometres. It causes sunburn, some forms of skin cancer, and is associated with eye problems such as cataract. Researchers working for the United States Environmental Protection Agency calculate that for every 1 per cent decrease in the concentration of ozone in the stratosphere there would be a 5 per cent increase in the number of non-malignant skin cancers each year.

More holes than one in the ozone layer?

The hole in the ozone layer over Antarctica was first noticed by researchers from the British Antarctic Survey, working at Halley Bay, in 1982. Ozone concentrations in the atmosphere overhead are measured using an instrument called a Dobson spectrophotometer, which analyses, in the spectrum of sunlight, the strength of the lines produced by ozone in the atmosphere. These measurements give an indication of the total amount of ozone in the column of air above the instrument. Scientists at Halley Bay have made such measurements regularly since the mid-1950s. They had never noticed before anything like the pattern that unfolded in 1982, and every year since.

Both spectrophotometer measurements and data from satellites in orbit around the Earth now show that each springtime in Antarctica (September and October) there is a massive depletion of ozone from the stratosphere. At the end of the long polar night, ozone is present in roughly the quantities that were normal in the 1960s and 1970s. But in a matter of a few weeks, at the start of spring, the overall concentration of ozone drops to about half its usual value. Observations from balloon and aircraft show that in some layers of the stratosphere, at an altitude of around 18 kilometres, virtually all of the ozone is destroyed during the Antarctic spring.

As the summer develops, the concentrations of ozone recover. Scientists base their explanations of what is happening on their findings from intensive studies of the Antarctic stratosphere since 1982, culminating in an international expedition by aircraft from Punta Arenas, on the southern tip of Chile, in 1987.

During the southern winter the air over Antarctica is cut off from the rest of the atmosphere by strong winds which sweep around the continent. These circumpolar winds act like a wall between the cold Antarctic air and the outside world. Behind that wall, the air gets so cold, down to –90 °C in the stratosphere, that clouds made of icy particles (polar stratospheric clouds) form. Chemical reactions take place on the surface of the particles involving chlorine compounds which are the pollution from human activities. These reactions release chlorine atoms. Thus, in contrast to the rest of the stratosphere, the majority of the chlorine will be converted from HCl to Cl_2 or possibly ClO. As soon as the Sun returns, chlorine in this form can be photolysed to release atomic chlorine, which participates in a series of reactions that destroy ozone. In one such chain of reactions:

$$ClO + HO_2 \rightarrow HOCl + O_2$$
$$ClO + ClO \rightarrow (ClO)_2$$
$$(ClO)_2 \xrightarrow{uv} ClOO + Cl$$
$$ClOO + M \rightarrow Cl + O_2 + M$$

The first steps occur even without sunlight; ClO and HOCl build up in the winter. The destruction begins with the arrival of solar ultraviolet radiation in springtime, which releases Cl to scavenge more molecules of ozone. The net effect is:

$$2O_3 \rightarrow 3O_2$$

and each chlorine atom can go round the cycle hundreds of thousands of times.

This, and similar cycles, explains the ozone depletion. With the return of summer, however, the polar stratospheric clouds evaporate and the chlorine is also eventually converted to other compounds, such as chlorine nitrate and HCl, until the following winter. The hole in the sky fills in, until the next spring. In October 1987, when the hole was at its biggest, it covered an area the size of continental United States, and took a slice out of the stratosphere as deep as Mount Everest is high. The American weather satellite Nimbus 7 recently provided scientists with measurements of atmospheric ozone which suggest that the ozone layer above the Arctic may also thin at certain times. The phenomenon does not seem, however, to be as dramatic as that above the Antarctic.

For now, the hole in the sky is an isolated, annual phenomenon, restricted to the polar regions. However, similar reactions to those that release chlorine in an active form from compounds such as chlorine nitrate, on the surface of icy particles in the polar stratospheric clouds, could take place on particles of dust from volcanic eruptions. In the past, major volcanic outbursts, such as the eruption of the Mexican volcano El Chichón, have had a noticeable effect in reducing the concentration of ozone in the stratosphere. The ozone layer may also be disturbed by changes in the Sun's activity. It is possible that, because there is more chlorine available each year as a result of human activities, future volcanic eruptions and natural solar variations will have a much greater effect on ozone, either locally (on the scale of a continent) or regionally (over a whole hemisphere).

Radiation with wavelengths in the band from 240 to 290 nanometres, known as UV-C, does not reach the ground at all today, so its effects on life are harder to predict. We cannot be certain, though, that under different conditions it might not reach the ground. In the laboratory, radiation within these wavelengths destroys nucleic acids (RNA and DNA) and protein – the basic molecules of life. A few plants have been tested for the effects of increased UV-B. Soyabean, a key crop in modern agriculture, suffers a 25 per cent decrease in yield when UV-B is increased by 25 per cent. Cattle, like people, are afflicted by eye complaints if UV-B is increased, including ailments known as cancer eye and pink eye.

The ozone layer is, clearly, a good thing for life on Earth. Just how much of this valuable material is there, shielding us from the harmful effects of solar ultraviolet radiation? In the whole stratosphere, at an altitude between 15 and 50 kilometres, there are only about 5000 million tonnes of ozone. If all the ozone could be brought down to sea level, and spread evenly around the globe, the pressure of the atmosphere above it

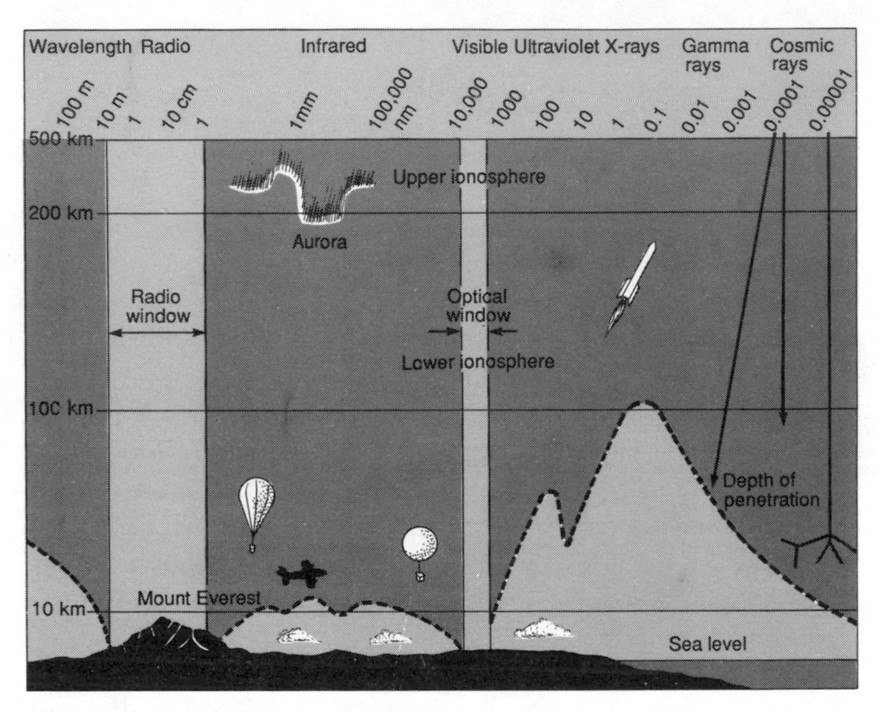

From the ground, optical and radio observations of the atmosphere are limited to two 'windows'. By raising instruments aboard balloons, rockets and satellites, scientists can now view the atmosphere across the entire spectrum

would squeeze it into a layer just 3 millimetres thick. A modest amount of gas, indeed, to do such an important job.

5 May 1988

Further Reading

Gaia by Jim Lovelock (Oxford University Press, 1982) explains the interactions between living things and the environment. *The Hole in the Sky* by John Gribbin (Corgi, 1988) gives an account of human influences on the ozone layer. *The Breathing Planet* edited by John Gribbin (Basil Blackwell, 1986) provides a guide to and through research on atmospheric sciences, culled from the pages of *New Scientist*. Several reports on stratospheric ozone are now available, including *Stratospheric Ozone* (HMSO, 1990) and *Into the Void?* (Friends of the Earth, 1987).

CHAPTER 14

The Greenhouse Effect

John Gribbin

The Earth is comfortably warm because it is surrounded by a blanket of air. 'Comfortably warm' means that the average temperatures on our globe are in the range between 0 °C and 100 °C where water is liquid. We can see that the atmosphere of the Earth has something to do with this by comparing the temperatures with which we are familiar with those on the surface of the Moon. The Moon is an airless 'planet' which is almost exactly the same distance from the Sun – the ultimate source of heat in our Solar System – as we are. On the airless Moon, the temperature rises to 100 °C on the sunlit surface, and falls to −150 °C at night. The surface temperature of the Moon averages about −18 °C. At this temperature the energy radiated from the Moon's surface into spacc just balances the incoming heat from the Sun. If the Earth had no blanket of air and was a bare, rocky ball like the Moon, it would also have an average temperature of about −18 °C. In fact, the average temperature of our planet in the layer of air just above the surface is about 15 °C. The blanket of air keeps our planet some 33 °C warmer than it would otherwise be.

How does it achieve this? **Solar energy** is radiated chiefly in the visible part of the spectrum, in a band from 0.4 to 0.7 micrometres. This radiation, and short-wave infrared, passes through the atmosphere of the Earth without being absorbed – although some of it is reflected back into space by clouds – and warms the surface of the land or sea. About 7 per cent of the Sun's energy is radiated at shorter wavelengths, below 0.4 micrometres, in the ultraviolet; this is important in maintaining a layer of ozone in the stratosphere (see Chapter 13). At the

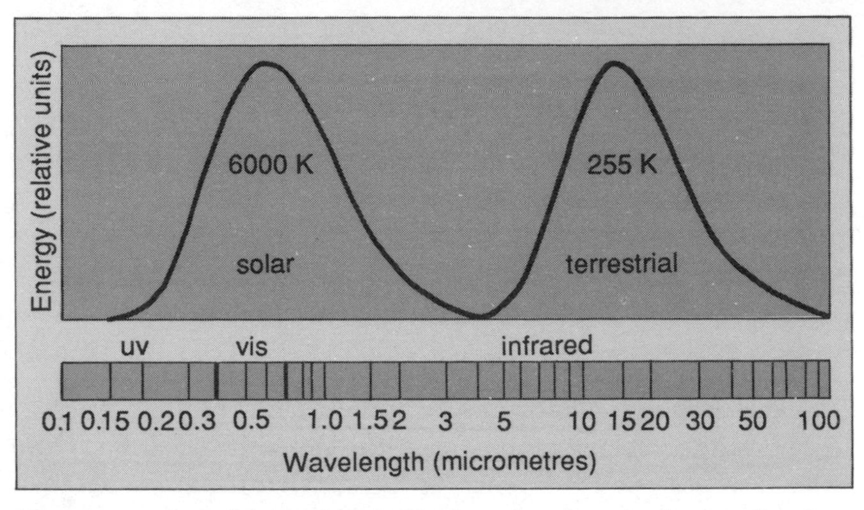

The same amount of energy that the Earth receives from the Sun is radiated back into space, but at longer wavelengths

other end of the spectrum, above 0.7 micrometres, energy is radiated in the infrared. This **infrared energy** is exactly the same as the radiant heat you feel when you warm your hands in front of a hot household radiator.

Incoming solar radiation
A blanket for the Earth

The wavelengths at which a hot object radiates most energy depend on the temperature of the hot object. The Sun has a surface temperature of about 6000 °C, and this corresponds to radiation in the visible band. The surface of the Earth, warmed by solar radiation, has a temperature of a few tens of degrees Celsius, and therefore radiates in the infrared, chiefly in the range from 4 to 100 micrometres. Water vapour absorbs strongly in the band from 4 to 7 micrometres, and carbon dioxide absorbs in the band from 13 to 19 micrometres. Between 7 and 13 micrometres there is a 'window' through which more than 70 per cent of the radiation from the Earth's surface ultimately escapes into space.

Because of the absorption, infrared heat radiated by the warm surface of the planet cannot escape freely into space, and warms the lower layer of the atmosphere – the **troposphere**. Because the air in the troposphere is warm, it radiates heat in turn, still at infrared wavelengths. Some of this heat goes back towards the ground, and keeps it warmer than it would otherwise be – the **greenhouse effect**.

As long as the amount of water vapour and carbon dioxide in the air stays the same, and as long as the amount of heat arriving from the Sun is constant, an equilibrium is established. The 'greenhouse gases' (carbon dioxide, water vapour and, as we shall see, one or two other gases) not only absorb but also emit infrared radiation. Because the temperature of the troposphere decreases with increasing height, the net effect is that each layer of the atmosphere absorbs energy radiated from below, at warmer levels, and passes it upwards, ultimately to escape into space at a lower temperature. The overall effect is to reduce the infrared radiated into space. The temperature at

Some of the energy radiated at infrared wavelengths from the ground is absorbed and re-radiated downwards by the atmosphere – the greenhouse effect

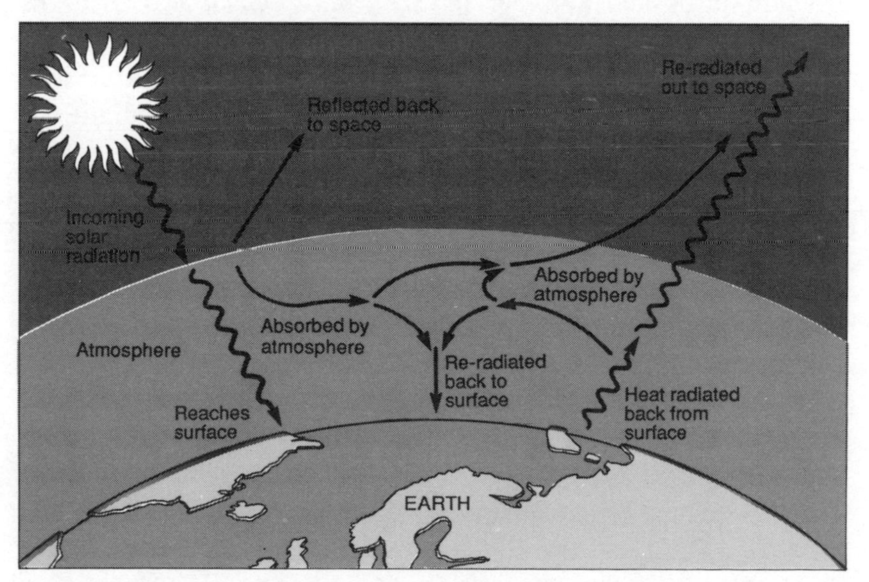

the surface must then rise until the amount of long-wave radiation leaving the Earth balances the amount of energy coming in from the Sun.

Both the ground and the air are warmed by the greenhouse effect. Present concern about the greenhouse effect began with the realization that human activities are upsetting the natural balance by increasing the amount of carbon dioxide in the atmosphere. This strengthens the greenhouse effect. It is now also clear that other gases being released as a result of human activities – other **anthropogenic gases** – absorb infrared radiation in the window from 7 to 13 micrometres where radiation used to escape freely. The combined effects of these anthropogenic greenhouse gases are likely to warm the world significantly over the next few decades.

The carbon dioxide trap
An enriched atmosphere

Scientists have known about the greenhouse effect since the middle of the nineteenth century. A British scientist, John Tyndall, published a paper in the *Philosophical Magazine* in 1863 about the effect of water vapour as a greenhouse gas. In the 1890s a Swede, Svante Arrhenius, and an American, P. C. Chamberlain, both considered the problems that might be caused by a build-up of carbon dioxide in the air and therefore a global warming, as a result of combustion of coal. Average surface air temperatures, worldwide, did indeed rise slightly in the early part of the twentieth century, by about 0.25 °C between 1880 and 1940. In the years following the US dustbowl of the 1930s, scientists suggested that this was a sign of the anthropogenic greenhouse effect at work. But between 1940 and 1970 the world cooled by about 0.2 °C, and the possibility of an increased greenhouse warming of the globe was not a major subject of scientific research. The situation changed because measurements of the amount of carbon dioxide in the atmosphere began to show a significant increase, stimulating a wave of scientific activity in the 1970s and forecasts that a doubling of the nineteenth-century 'natural' concentration of carbon dioxide would cause the world to warm by about 2 °C.

Between 1970 and 1980, just at the time these forecasts were being made, the world's mean temperature increased by about 0.3 °C, and the rapid warming seems to have continued into the 1980s. Although records of temperature go back only to the 1850s, 1989 was the warmest year on record. There is no proof that this is solely, or even primarily, a result of the anthropogenic greenhouse effect. But it certainly gives research into the greenhouse effect a topical interest.

Accurate measurements of the amount of carbon dioxide in the air began to be taken at Mauna Loa (in Hawaii) and at the South Pole, in the International Geophysical Year, in 1957–8. These measurements are an important guide because they are taken far away from any major sources of industrial pollution,

Annual average temperatures since 1880. (Source: J. E. Hansen and S. Lebedeff)

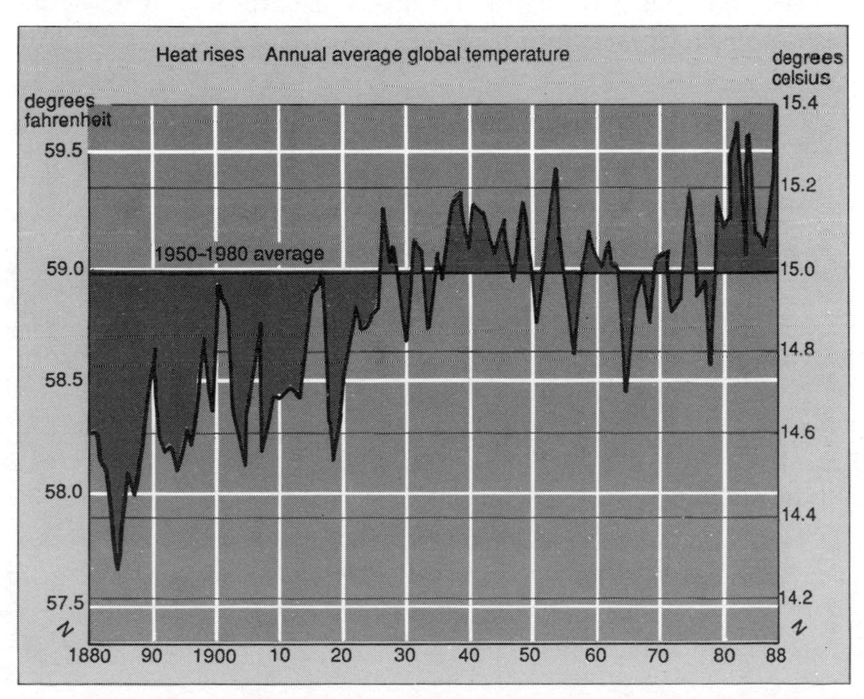

and represent the 'well-mixed' state of the atmosphere. They show a clear annual rhythm, associated with the seasonal changes in vegetation over the land masses of the Northern Hemisphere – the Earth's vegetation 'breathes' carbon dioxide in and out over an annual cycle (this is dominated by the Northern Hemisphere because that is where most land lies). But by the 1970s it was also clear that this annual cycle is superimposed on a rising trend of global mean carbon dioxide concentration.

In 1957 the concentration of carbon dioxide in the atmosphere was 315 parts per million (ppm). It is now about 350 ppm (0.035 per cent). Most of the extra carbon comes from burning fossil fuel, especially coal; part of the increase may be due to the destruction of tropical forests. When 1 tonne of carbon, perhaps in the form of coal, is burnt, it produces about 4 tonnes of carbon dioxide, as each carbon atom is combined with two atoms of oxygen from the air. By the early 1980s some 5 gigatonnes (5000 million tonnes; 5 Gt) of fuel was being burnt each year, so the annual input of carbon dioxide to the atmosphere from combustion of fossil fuel had reached about

The build-up of carbon dioxide in the air, recorded at Mauna Loa Observatory, Hawaii

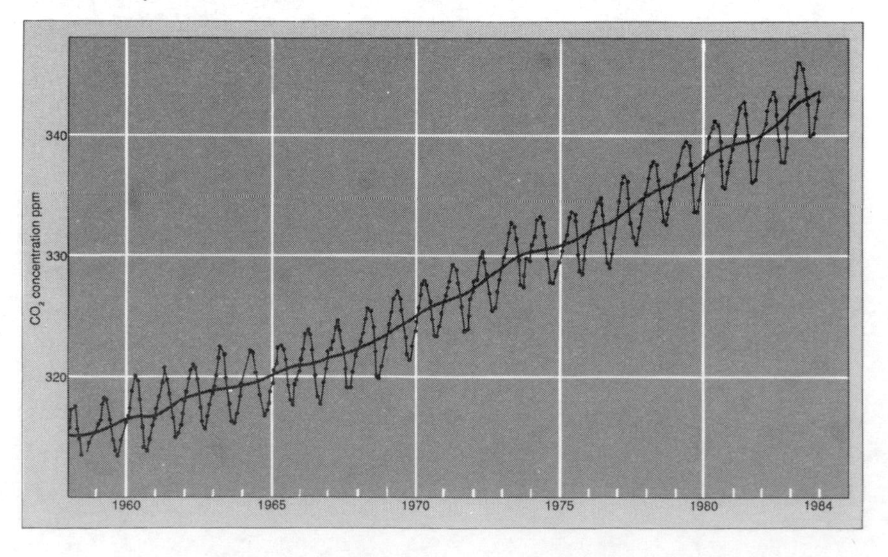

20 Gt. The increase in the carbon dioxide concentration of the atmosphere, however, corresponds to slightly less than half the amount of carbon dioxide produced by human activities each year. Roughly half the carbon dioxide we produce is absorbed in some natural sink (or sinks). Some may be taken up by vegetation, which grows more vigorously in an atmosphere enriched with carbon dioxide – 'vegetation' here includes the biomass of the sea, dominated by microorganisms such as plankton. Some may be dissolved in the oceans.

Between 1850 and 1950 roughly 60 Gt of carbon was burnt, chiefly as coal, in the burgeoning Industrial Revolution. As much carbon again is now being burnt every dozen years. The input of carbon dioxide to the atmosphere between now and the end of this century will probably be as great as the input in the hundred years from 1850 to 1950.

Researchers estimate that in the middle of the nineteenth century the natural concentration of carbon dioxide in the atmosphere was about 270 ppm. They have also measured, in bubbles of atmospheric gases which have been trapped in the polar ice sheets, similar concentrations of carbon dioxide from the period before industrialization. Calculations of the likely effect of the build-up of carbon dioxide in the atmosphere on climate are usually presented in terms of the increase compared with the background level of 270 ppm. Studies of air bubbles trapped in ice cores from the Antarctic show that roughly the same concentration of carbon dioxide persisted throughout the past 10 000 years.

Climatic changes
Hotter, wetter weather

Different computer models of the global climate produce slightly different figures, but it is a reasonable rule of thumb that doubling the carbon dioxide concentration of the atmosphere from the baseline of 270 ppm will produce an increase in global mean temperatures of about 2 °C. But this is

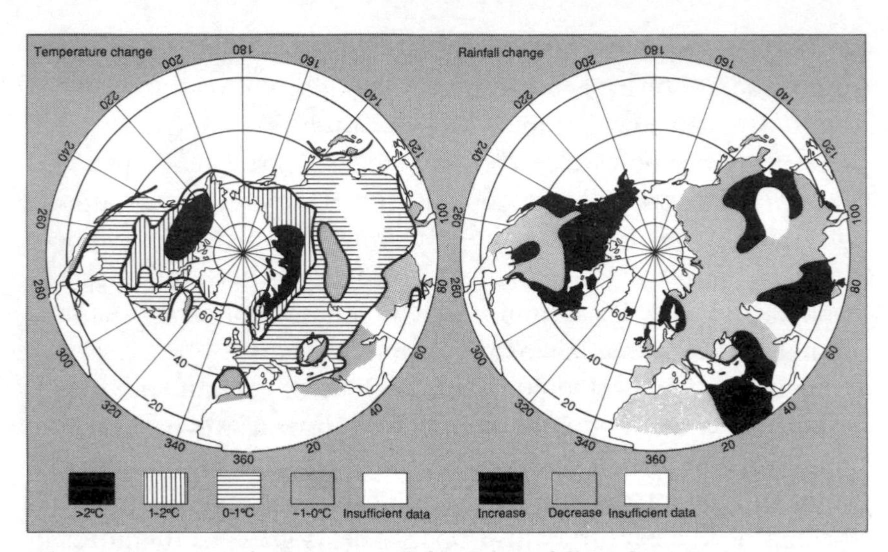

Changing patterns of temperature (left) and rainfall (right) in a warmer world. (Source: Climatic Research Unit, University of East Anglia)

less than half the story. The same computer models show that the warming is likely to be much greater at high latitudes, near the poles, while the tropics, which are already hot, will warm only slightly. At the same time, there will be changes in prevailing winds and in the distribution of precipitation around the globe. One key forecast is that continental interiors will dry out as the greenhouse effect takes a grip.

Simply extrapolating the present increase suggests that doubling of the carbon dioxide concentration will occur in almost exactly 100 years from now, in the 2080s. This takes no account of the ways in which the use of energy globally is likely to change over the next century; the matter is now the subject of eager debate. But we can use this figure to put the carbon dioxide 'problem' into perspective alongside the greenhouse effect of other gases being released to the atmosphere.

Several other gases that are being released by human activities – such as ozone, methane, nitrous oxide and chlorofluorocarbons (CFCs) – absorb infrared radiation in the window from 7 to 13 micrometres. CFCs, best known as the chemicals responsible for the holes in the ozone layer over the Arctic and Antarctica (see Chapter 13), are very efficient greenhouse gases. One molecule of either of the two most common CFCs has the same greenhouse warming effect as 10 000 molecules of carbon dioxide.

Methane, which at present has an atmospheric concentration of 1.7 ppm, is increasing at a rate of 1.2 per cent per year, probably because of the biological activity of bacteria in paddy fields, and also because of the release of natural gas from commercial oil and gas fields. Nitrous oxide is building up from a concentration of 0.3 ppm at a rate of 0.3 per cent a year as the use of nitrogen-based fertilizers increases. Ozone near the ground (in the troposphere) is increasing as a result of human activities.

Veerhabadrhan Ramanathan, of the University of Chicago, has converted the greenhouse effects of all these gases into carbon dioxide equivalents, and projected their growth forward to the year 2030. By then, he claims, these minor contributions will probably add up to as big an effect as that of the extra carbon dioxide produced by human activities. In effect, they will double the strength of the anthropogenic greenhouse effect. On this basis, the *effective* doubling of the background concentration of carbon dioxide will occur by about the year 2030, half a century sooner than if carbon dioxide alone were increasing.

Climatic conflict
Greenhouse *v.* ice age

How will this affect the weather of the world? Nobody can say exactly, but one way to get an idea of the direction in which things will change in different regions of the globe is to look at the difference in weather patterns for warm years and cold years during the present century. Researchers at the Climatic Research Unit, at the University of East Anglia, carried out just such a study. They used data from the 50-year interval 1925–74, and picked out the five warmest and five coldest years during that interval. The team looked first at temperature records from around the Arctic, between 65 °N and 80 °N. This includes northern Norway, northern Sweden, much of Finland, Iceland, northern Canada, Alaska and northern Siberia. In this zone the temperature difference between the warm and cold extremes is 1.6 °C, but the temperature difference averaged over the whole Northern Hemisphere is only 0.6 °C. Comparing only winters, the warm extreme was 1.8 °C warmer at these latitudes, while in summer the temperature difference was only 0.7 °C. Overall, there is a 1–2 per cent increase in rainfall in the warm years, which is expected since more water evaporates from the ocean when the world is warmer. But this modest average increase conceals much bigger regional effects, including a *decrease* in rainfall over much of the US, Europe, the USSR and Japan. India and the Middle East have a greater than average increase in rainfall in this scenario.

This single study cannot be taken as a definitive guide to how weather patterns will change as the world warms, but it bears out the computer forecasts (suggesting that high latitudes warm three or four times more than average). It also indicates that the patterns we are all used to, and on which present agricultural practices are based, are unlikely to be 'normal' in the years ahead.

In itself, the greenhouse effect is not necessarily a bad thing. It is almost certainly a more desirable state of the world, from the human point of view, than the onset of a new ice age might be. But what matters is the way human society responds to this environmental challenge. It is now certain that the world will change, as a result of the greenhouse effect, during the lives of all of us. Plans that are made on the basis of how things used to be – agricultural plans, designs of flood barriers, calculations of the need for reservoirs of drinking water, and so on – will almost certainly be wrong. This is why climatologists now point to the example of the droughts which hit the US in 1988. Those droughts were not caused directly by the greenhouse effect. Although 1988 is the warmest year on record, 1987 was nearly as warm and there were no comparable droughts then. But *whatever* the immediate cause of those droughts, they are an indication of the kind of shortages of rainfall that are likely to be increasingly common in the heart of North America, and other continents, as we move into the twenty-first century.

Rising sea levels and melting ice caps

The other immediate effect of an increase in global mean temperatures is a rise in sea level. This is already happening. Sea level has risen by about 15 centimetres during the twentieth century, and the rise is very much in line with the rise in temperature that has occurred over the same interval. Most of this rise can be explained simply in terms of the thermal expansion of sea water. Only a little extra water has been added to the sea, by melting glaciers on mountains at low latitudes. Paradoxical though it may seem, at present the polar ice caps may be *increasing* in size. This is because more moisture is evaporating at low latitudes, and this is falling as snow near the poles, where it is still cold. A global warming of about 2 °C, possible within 40 years, will increase sea level by a further 30 centimetres or so, largely because of the expansion of sea water.

But one 'scare story' associated with the greenhouse effect is dismissed by the experts. This is the fear that the entire West Antarctic ice sheet might collapse, sliding into the ocean and raising sea level worldwide. Some calculations do suggest that once the world warms by about 4 °C (which could happen before the next 100 years is over) the ice sheet might 'collapse'. But what glaciologists mean by a collapse is still a slow process by everyday standards – it would take several hundred years for all the West Antarctic ice to slide into the sea, eventually raising sea levels by 5 metres or more, but only at a rate of one or two centimetres a year. There would be ample time to walk out of harm's way, although the impact on coastal cities and low-lying countries, such as Bangladesh and the Netherlands, would be catastrophic in the long term.

The greenhouse effect should be a factor in all long-term planning, and it impinges directly on politically contentious issues such as the choice between nuclear and fossil fuel for future generations of power stations. Ironically, in this context it is nuclear power that is the environmentally clean alternative.

That example highlights the difficulty of coming to grips with the greenhouse effect.

22 October 1988

Further Reading

The definitive study is *The Greenhouse Effect, Climatic Change and Ecosystems* edited by Bert Bolin, Bo Döös, Jill Jäger and Richard Warrick (Wiley, 1986). It is big and expensive, but worth investigating in a library. *Hothouse Earth* by John Gribbin (Black Swan, 1990) puts the greenhouse effect into perspective against the history of global climate, and the *New Scientist* guide *The Breathing Planet* (Basil Blackwell, 1986) goes into a little more detail.

CHAPTER 15

Plants, Water and Climate

Ian Woodward

Liquid water is a vital ingredient for all living organisms. Up to 90 per cent of the mass of plants and animals is water. Without water a plant wilts, and eventually dies. Without plants almost all other forms of life would die. This makes the process by which water moves from the soil into the plant one of the most crucial of all.

The movement of water from the soil through the plant is complex and depends both on factors that the plant can control and some that it cannot – such as exposure and temperature. Some forms of human activity, agriculture and industry for instance, can affect those factors that are outside a plant's control, and so interfere with the normal flow of water from the soil to the air, via the plant. In moist, temperate climates, for example, felling a forest and replacing it with grassland can cut by half the amount of water that evaporates from the vegetation covering that area of land. As a result, more liquid water flows from the land into streams. This lost water carries away nutrients vital for the growth of plants.

Water can exist in three phases: solid ice, liquid water and gaseous water vapour. Water is a liquid at the temperature at which plants grow only because its molecules are linked by **hydrogen bonds**, which join the hydrogen and oxygen atoms of adjacent water molecules. Methane and carbon dioxide, which have similar molecular weights to water but no hydrogen bonds, are gases at this temperature. Hydrogen bonds are weak. They have only 4 per cent of the strength of a covalent chemical bond, such as between hydrogen and oxygen in a single molecule of water. Yet this strength is perfect for developing a range of unusual properties, which are vitally important to living organisms. For example, when water evaporates it changes from a liquid to a gas. Before a water molecule can escape from the liquid, it must break its links with other water molecules. Energy is needed to break the hydrogen bonds – 2450 joules of energy evaporates 1 gram of water at 20 °C. This energy comes from heat stored in the liquid water. Because of this, evaporation, or **transpiration** as it is called when water evaporates from a leaf, is an effective way to cool a plant. On a dry, bright summer's day the energy extracted from the leaves of a large rose bush during transpiration is equivalent to about one quarter of the energy emitted by a 1-kilowatt electric fire.

Every cell in a plant that is actively growing must have water, so the plant must have an efficient system for transporting water to the farthest leaf and shoot. Water travels the greatest distances in the plumbing system, a network of tubes called the **xylem**. Hydrogen bonds play a vital part in the transport of water through the xylem. Bonds between water molecules and the walls of the xylem tubes help the upward movement of water in the xylem. This bonding also holds the water together, so that it can withstand stretching as it is pulled up the tube. If the column of water is not strong enough and breaks under tension, bubbles of air enter the xylem pipes and block any further movement of water. The giant redwood, *Sequoia sempervirens*, can grow as tall as 100 metres, and leaves at the top of the tree require a continuous supply of water to replenish what they lose by transpiration. The water column is at least 100 metres long. To ensure that water reaches the leaves, the difference in pressure between the bottom and the top of the tree must be at least enough to overcome the gravitational pull on the column of water. This is equivalent to the pressure required to lift a weight of 1.6 tonnes resting on the palm of the hand.

The living plant cell is encased in a thick cellulose wall, which can withstand pressures of at least 1 megapascal (10

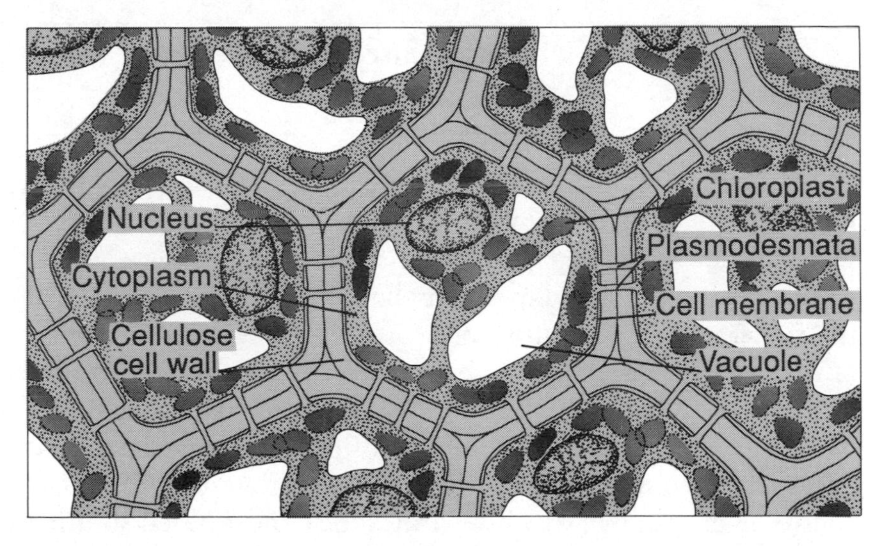

Water must reach every cell. It can travel speedily through the fibres of the cell wall, or slowly via fine strands of cytoplasm, called plasmodesmata

times atmospheric pressure). Inside the cell wall, the cell is mostly watery jelly, the **cytoplasm**, enclosed by a thin plasma membrane. Again, the unique properties of water make possible the biochemical and biophysical activities of the cell. Water needs to absorb or lose a lot of heat before its temperature changes significantly – that is, it has a high specific heat capacity. This property protects the cell from excessive changes in temperature. Water is a very effective solvent for many compounds and biochemical reactions take place in aqueous solution. Water is also transparent, so sunlight can penetrate the cell and reach the **chloroplast**, where the light energy drives **photosynthesis**, the process by which plants manufacture carbohydrates.

The biochemical processes that take place in the cell, including photosynthesis, produce solutes, such as sugars, which dissolve in the cell's water. Some of the hydrogen bonds in the water then connect to these solute molecules rather than to water molecules. The solution outside the cell is usually weaker than that inside the cell; it has a lower concentration of solute molecules and a higher concentration of water molecules. Fewer hydrogen bonds are tied to solute molecules and so the water has more internal energy.

Cells with potential
Osmosis and turgor

The difference in the energy in pure water and in a solution with water can be used to do work, such as the movement of water – but only in a particular set of circumstances. Pure water flowing out of a tap under the force of gravity moves at the same rate as a solution of sucrose flowing from a similar tap. The difference in energy between them has no effect on the rate at which they flow. But a plant can take advantage of the energy difference between solutions of different concentrations because of the nature of the plasma membrane inside the cell wall. The membrane, which separates the concentrated solution inside the cell from the more dilute solution in the cell wall, is permeable to water molecules but not to solute molecules. It is said to be **semi-permeable**. This property allows the cell to use some of the difference in energy between the two solutions to force water into the cell. This mass flow of water is called **osmosis**.

The movement of water into a cell by osmosis should cause the cell to swell, increasing its volume. Although the cell swells a little, the strong cellulose cell wall resists the expansion, so putting pressure on the contents of the cell. This pressure opposes the osmotic flow and increases the energy of the water in the cell. Pressurization continues until no more water flows into the cell. At this point the energy levels inside and outside the cell are equal, a state defined as:

$$\Psi_c = \Psi_p + \Psi_s$$

where Ψ_c is the cell **water potential**, Ψ_p is the degree of pressurization within the cell, or **turgor potential**, and Ψ_s is the **osmotic potential**. There is no net flow of water when Ψ_c is equal to the water potential outside the cell.

Osmotic potential is a measure of how much internal energy is lost when a solute is added to pure water. As the water potential of pure water is zero, osmotic potential must be negative. Water which is under pressure – restrained by the cell wall – has a greater internal energy than pure water at atmospheric pressure and so has a positive water potential.

A growing cell must have a high turgor potential – that is, its cytoplasm must exert a high pressure – to force the immature cell wall to expand and stay expanded. If the cell is short of water, it cannot maintain its turgor. If the turgor potential falls to zero, the cell wilts and the lack of water may inhibit many biochemical processes. Growing cells operate best when turgor potential is high. Each cell must draw the water it needs to maintain its turgor potential from the 'mains supply' as it flows through the plant. A cell can tap into two interconnecting waterways, called the **apoplast** and the **symplast**. The apoplast is the route along interconnecting cellulose cell walls, in the spaces between the cellulose fibres. Water travels fastest by this path. The symplast is a slower route, through individual cells, by way of fine strands of cytoplasm, or **plasmodesmata**, which pass through the cellulose walls, and connect the living parts of the cells. These two pathways operate throughout the plant from the **root hair** that has the very first contact with water in the soil to the **mesophyll cells** of the leaf, where water becomes vapour and leaves the plant.

Cells at the surface of a root take in water by osmosis. Any cell can draw in water but those near the tip of the root and in the zone of root hairs are much more efficient at it. Root hairs are small protuberances from cells on the surface of the root. They penetrate between soil particles and make intimate contact with water in the soil. These cells take up water as much as six times as fast as cells in older regions of the root, which are coated with a water-repellent called **suberin**. The suberin prevents water from moving out of a root when the soil is dry.

Ways to the top
From root to leaf

Once taken into the surface cells, the water moves immediately into the apoplastic and symplastic pathways. Most water flows

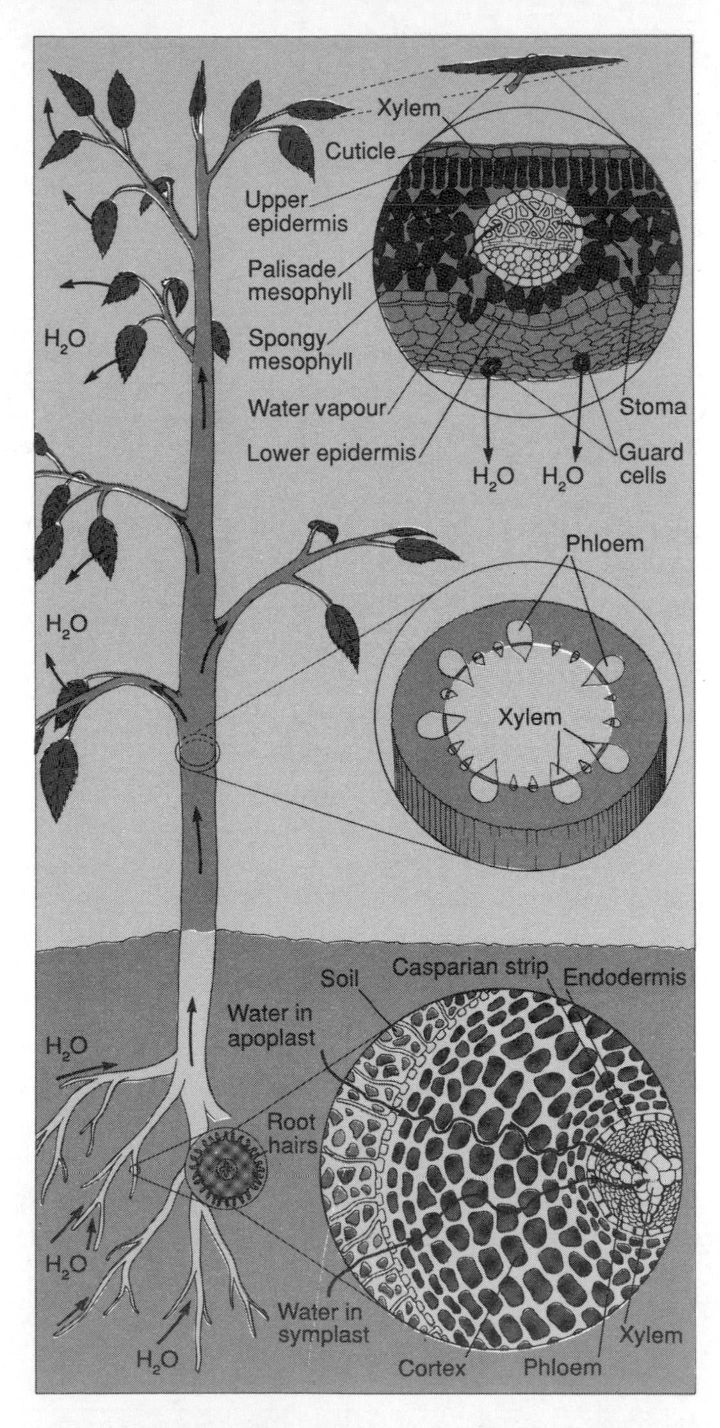

The world's most important waterway: water from the soil travels from the root hair to the xylem. Once in the xylem, it moves swiftly to every shoot, finally leaving through pores in the leaves

quickly through the apoplast until it reaches a ring of cells around the xylem at the centre of the root. This ring of cells forms the **endodermis**. At the endodermis, the way through the apoplast waterway is barred by strips of water-repellent suberin, called **Casparian strips**, in the end walls of the endodermal cells. The water is forced to take the alternative route through the symplast waterway and into the xylem. The endodermis acts as a controlling valve between the living and fragile cells of the root and the dead and very strong cells of the xylem.

Xylem cells may be 10 or 20 times as long as other cells in the root; they are empty, with no cytoplasm or nucleus, and serve purely as pipes for carrying water. The structure of xylem cells varies with species, but in most cases the ends of the cells are pitted and porous. Such a structure allows water to move as a continuous column from the endodermis of the root to the xylem cells that make up the veins in the leaf.

For plants in sunlight there is a gradient of water potential – and so also in the amount of energy available for work – from the highest in the soil to the lowest at the leaf. This gradient is the source of the energy that forces the water upwards to the leaf. The gradient of water potential between the bottom and the top of the xylem may be as much as 1 to 2 megapascals (10 to 20 times atmospheric pressure). The rate at which water flows through the xylem is determined by the drop in energy from one end of the xylem to the other; by the diameters of the xylem cells; and by how porous their end walls are. Water meets a certain amount of resistance from the walls of the xylem vessels and the pitted ends of the cells. Xylem cells with the greatest diameters offer the least resistance to the flow of water. Vines and the tangled climbers of the rainforest, lianas, have xylem cells up to a millimetre across, and water courses through them at a rate of about 400 millimetres a second. At the other extreme, the xylem vessels of conifers may be only 0.01 millimetres across, and water moves at a sluggish 0.5 millimetres a second.

As a plant grows, the xylem becomes longer, so increasing

the resistance to the flow of water, and the energy needed to overcome it. Tall trees also need significant amounts of energy to move water against the force of gravity. The drop in water potential across the xylem is a measure of the energy needed to overcome these forces. To overcome gravity, the drop is 0.01 megapascals per metre of height. So for a tree 100 metres tall the drop in water potential will be 1 megapascal. Additional energy is required to push water through the end walls of the xylem cells and against the frictional resistance of the xylem walls. For a tree 100 metres tall this would amount to a drop in water potential of another 1 megapascal.

The total drop in water potential of 2 megapascals is close to the limits of the cohesive strength of the continuous column of water in the xylem. If the drop in water potential is any larger, the water column may break, allowing air to enter the water column through the walls of the xylem. This rupture of the water column, known as **cavitation**, prevents the flow of water through the xylem cell where the air bubble forms. The problem is exactly akin to the problems of air bubbles in the hydraulic braking system of a car. In such a case, the driver may depress the brake pedal completely without transferring enough pressure to operate the brake mechanism. Cavitation happens regularly. Perennial plants get around the problem of airlocked vessels by growing new vessels each year to ensure a continuous flow of water. In very long-lived plants, such as trees, a large proportion of the xylem cells no longer function because of cavitation. Ultimately, cavitation probably limits the maximum height of a tree.

Once water reaches the xylem vessels of the leaf, it moves through the two waterways, the symplast and the apoplast, to cells that line the air chambers beneath the pores, or **stomata**, in the epidermis of the leaf. Most of the water that the plant loses by evaporation escapes from these surfaces. Because water leaves these surfaces faster than it can be supplied, these cells have the lowest water potentials of the whole route from soil to leaf. This ensures that the gradient is maintained and water moves continuously through the plant. At the surface of the

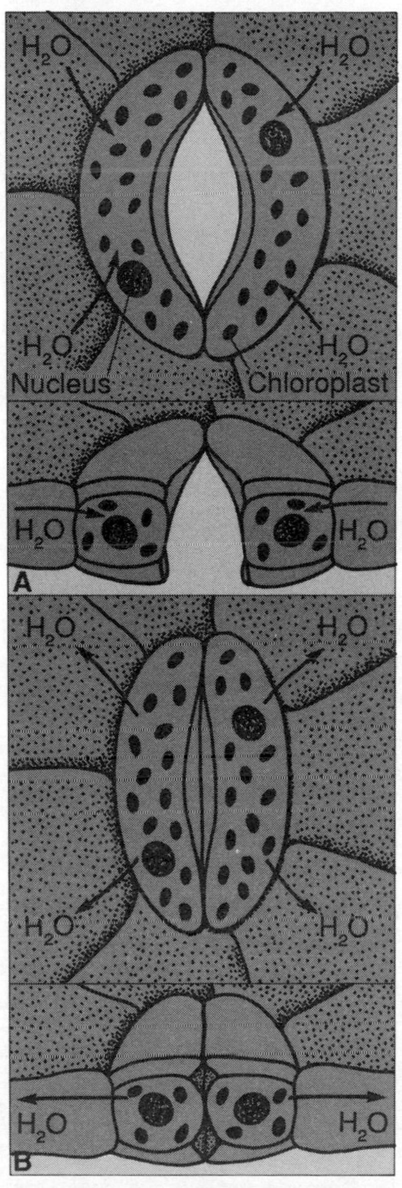

Stomata open when water moves into the guard cells and makes them turgid. The thick inner sides of the guard cells bow outward, opening the pore (A). The guard cells maintain their turgor by accumulating solutes, forcing water to move in by osmosis. When the concentration of solutes is low, water moves out into the surrounding cells and the guard cells begin to collapse, closing the pore (B). Plants control the stomata by regulating the movement of ions into and out of the guard cell

mesophyll cells, water changes phase from liquid to vapour by extracting the energy for vaporization from the leaf.

During the day, stomata open in response to solar radiation. Carbon dioxide diffuses through the stomata and into the leaf. From there it diffuses to the chloroplasts, where it is converted to carbohydrates in the process of photosynthesis. At the same time as carbon dioxide enters the stomata, water vapour diffuses out to the drier air around the leaf. This unavoidable loss of water vapour starts up the stream of water from the soil through the plant, to the sites of evaporation.

The plant can control the rate of transpiration to some degree by varying the extent to which it opens its stomata. Each pore is surrounded by a pair of specialized epidermal cells called **guard cells**. Small changes in the water potential of the guard cells cause them to change shape, opening or closing the stomata. When the pores are closed, water vapour may still escape by diffusing through the epidermis of the leaf, but this is kept to a minimum because epidermal cells are coated with **cutin**, a water-repellent wax. In sunlight the plant keeps a tight control on how far its stomata open, responding to the dryness of the air and the water potential of the leaf.

Outward bound
From leaf to air

Once the water vapour emerges from the stomata, it is lost to the air. It does not move away immediately, however, but lingers in a slow-moving layer of air known as the **boundary layer**, which surrounds the leaf. The flow of air in the boundary layer is sluggish because of friction presented by the leaf. A large rhubarb leaf offers considerable resistance to the flow of wind and may have a deep boundary layer, perhaps several millimetres thick. The needle-leaf of a conifer exerts little frictional force against wind and has a thin boundary layer, in the order of fractions of a millimetre. The diffusion of water vapour and the loss of heat through the boundary layer may be some 1000 times slower than in the turbulent air farther from the leaf. In bright sunshine the rhubarb leaf may be as much as 10 to 15 °C warmer than the needle-leaf, even though they have the same number of stomata open.

The depth of the boundary layer is important in determining the rate of transpiration. With leaves at the same temperature, the thicker the boundary layer the more it slows the loss of water. But the depth of the boundary layer changes with the speed of the wind. As the wind grows stronger the boundary layer becomes thinner; water escapes from the layer faster and the rate of transpiration increases. The increase might be enough to cause the water potential in the leaf to fall – when the stomata close.

The movement of water from the soil to the plant is a vital and complex process. How far a plant can control it depends to a large extent on its environment. Changes to that environment can disrupt the process at almost any stage.

Changes on the way
Climate and clearances

Modern agriculture has changed the face of the landscape in most of the developed parts of the world, replacing forests and complex habitats with cereals and pastures. The boundary layer of grassland is much thicker than that of forest because the air over grass tends to be stiller, allowing a deeper layer of air to lie unstirred. In bright sunlight the presence of a deep boundary layer causes the temperature of the grassland to increase, until it is much higher than that of both the air above and the forest. The higher the temperature the greater the potential for transpiration. If the stomata of the grass leaves close partially, both the temperature and the potential for transpiration increase still more – so cancelling out the antitranspirant effect of closing the stomata. In the grassland canopy, then, the thick boundary layer exerts stronger control over transpiration than the stomata do, at least until the stomata are tightly closed.

In forests the boundary layer is much thinner and so the stomata have more control over transpiration and the temperature of the canopy is always close to air temperature. Drought is much more likely where the land is covered by pasture.

Human activities that affect transpiration are not restricted to alterations to the vegetation. By steadily adding carbon dioxide and other greenhouse gases to the atmosphere, we are also changing the world's climate. Within decades the Earth will be warmer and drier (see Chapter 14).

How these changes will alter the flow of water through the world's vegetation, and what the consequences will be, is examined in Chapter 16.

18 February 1989

Further Reading

Water Flow in Plants by John Milburn (Longman, 1979) provides an advanced treatment (A-level standard) on this theme.

CHAPTER 16

Plants in the Greenhouse World

Ian Woodward

A world in the grip of the greenhouse effect will be a very different place from the one we know today. By the middle of the next century the climate will be much warmer, perhaps as much as 12 °C warmer near the poles. Patterns of rainfall will change: some places will be wetter, others drier. The amount of carbon dioxide – one of the gases responsible for the greenhouse effect – will continue to build up in the atmosphere (see Chapter 14). All these factors affect the way a plant draws up water through its roots. They therefore determine which types of plant grow in particular places. One of the most striking features of the greenhouse world will be the new look of the belts of vegetation that girdle the globe.

Plants draw water from the soil and distribute it to every shoot and leaf. The plant has a good deal of control over the movement of water and increases or restricts the loss of water to the air by opening or closing the pores in its leaves (see Chapter 15). These mechanisms must operate very efficiently in a healthy, growing plant, for if the amount of water in the plant's tissues falls by just 5 per cent, its growth begins to slow down; if the amount decreases a further 10 per cent, **photosynthesis** stops. Photosynthesis is the process by which a plant converts carbon dioxide and water to carbohydrates, using energy from sunlight. So if a plant is to carry on growing, it must keep a tight control on the amount of water it loses.

If we scale up our outlook from an individual plant to a swathe of vegetation, we can see the same degree of sensitivity.

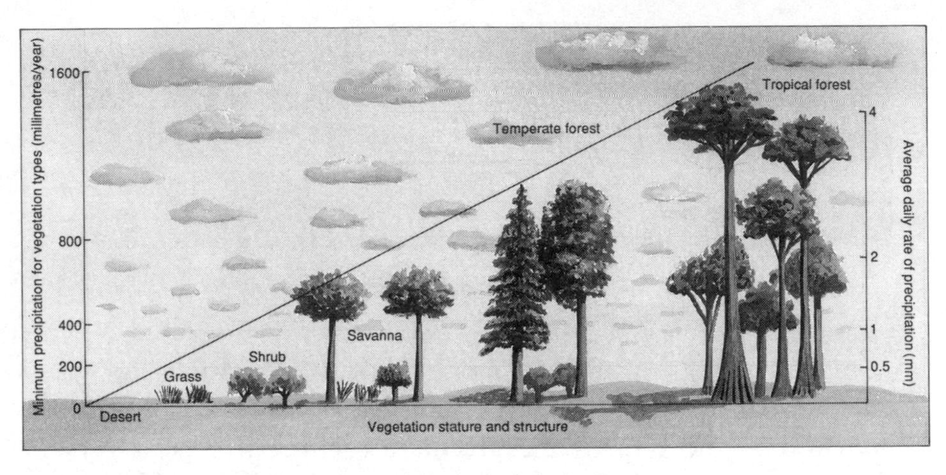

Life support: the higher the rainfall, the lusher the forest

Where water is plentiful the vegetation is dense and lush; where it is in short supply plants are sparser. The natural vegetation in places that have consistently heavy rain (more than 1200 millimetres a year) is a tall, dense forest. Places that receive less than 200 millimetres of rain (or snow) in a year have a totally different sort of vegetation, made up of sparse short individuals, perhaps forming a grassland.

Vegetation is a reservoir of water; each type has a different capacity for storage. The trunks and stems of the trees in a dense tropical forest hold the equivalent of about 25 millimetres of rainfall. A grassland holds only a fraction of this amount – perhaps equivalent to about 1 millimetre of rain. When the soil is wet the amount of water stored in the trunks of forest trees remains constant from day to day, though it falls a little in the early afternoons. This is the time when the rate of evaporation of water through the pores of the leaves, or **transpiration**, is highest. When the soil begins to dry out, transpiration carries on in order to maintain the flow of water to the leaves. To continue doing this, the plant draws increasingly on the water in its trunk. Because trees in a tropical forest can hold so much water in reserve, they can continue to supply water to the leaves, and so maintain growth and photosynthesis, for four or five days during dry periods, even though the rate of transpiration is high. Shrubs have smaller reserves of water, enough to maintain a supply to their leaves for only half a day.

If it rains during the period when the plant is tapping its own reserves of water, the plant can draw in more water from the soil and refill the reservoir. The plant will continue to grow. If the water shortage continues, perhaps through a dry season, then the plant will limit, or even stop, some of its activities, such as photosynthesis and growth; this shutdown may last one or two months. If the drought lasts longer, the plant may have to take more drastic action, and shed its leaves. By casting off its leaves, a process called **abscission**, the plant prevents further loss of water. This preserves water in the tissues that are crucial for regenerating leaves when conditions improve. These vital tissues are the dormant buds and **meristems**, undifferentiated tissues from which new leaves and vascular tissue develop. If the drought continues, however, the plant will die.

It follows that the natural vegetation growing in a particular climate is able to withstand the driest period of the year, and

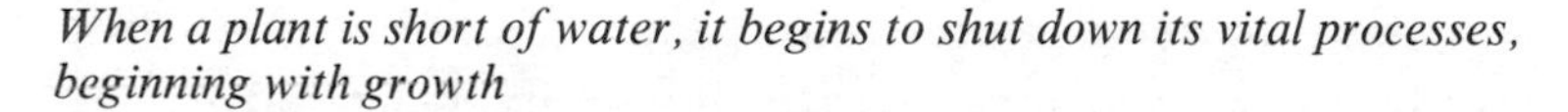

When a plant is short of water, it begins to shut down its vital processes, beginning with growth

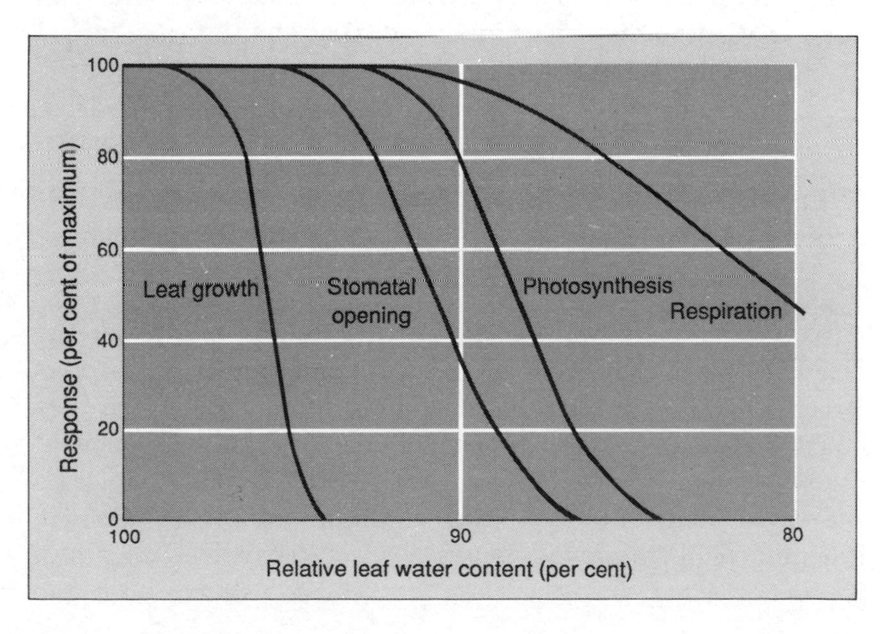

has the capacity to grow quickly when water is abundant. This close tie between vegetation and rainfall allows botanists to make broad predictions of the vegetation they will find in any area of the world, on the basis of the amount and seasonality of rain. In general, the stature and complexity of vegetation increases with precipitation.

Predictions based on rainfall alone provide a broad outline of the vegetation but little detail. Temperature, especially the lowest temperature in winter, is also important to a plant. Plants are 90 per cent water and may freeze when the temperature falls below 0 °C. If the water inside a cell freezes, the cell dies. The plant can tolerate freezing of water outside the cell, in the latticework of the cellulose cell wall and the spaces between cells. The plant protects itself to some extent by moving water from inside its cells to the intercellular spaces and the cell wall. This movement prevents the living part of the cell from freezing but leads to a shortage of water in the cell, a state known as **frost drought**. The plant is further protected by a physical process called **supercooling**, in which the freezing point of water is lowered. Pure water usually freezes at 0 °C; but if it cools slowly at the sort of rate that is likely in nature, at about 1 °C per hour, then it may not freeze until it has cooled to −2 or −3 °C. The presence of solutes dissolved in the water, as in plant tissues, can increase this capacity for supercooling.

Limits of tolerance
What grows where

Just how far plants can tolerate frost drought and lower the freezing point of their tissues determines where they can live. Broad-leaved evergreen vegetation stretches from the hot equatorial region to the Mediterranean region. Trees such as the holm oak, which grow in a Mediterranean climate, tolerate the lowest temperatures – to a minimum of −15 °C. Where minimum temperature falls below −15 °C, the typical vegetation is broad-leaved but loses its leaves in winter, like the pedunculate oak. These trees dispense with their leaves in the autumn, after extracting useful nutrients and storing them in the overwintering meristems and buds. These two tissues must be resistant to frost, but there is a limit to their tolerance. Most broad-leaved deciduous trees can survive to a temperature of between −40 and −50 °C. Where temperatures fall below −50 °C, broad-leaved trees give way to pines, firs and spruces, the needle-leaved conifers that make up the vast belts of boreal forest. A few broad-leaved deciduous species of birch and poplar cling on, but only where they are exposed to the warmth of the Sun. Botanists do not understand exactly how these northern trees prevent their overwintering buds and leaves from freezing.

With information on both rainfall and temperature, predictions of vegetation become more precise. With a third piece of information, the picture becomes clearer still. The growing season is a vital time for a plant. The species that make up particular types of vegetation need a minimum length and warmth of growing season to complete all the processes of their life cycle. This combination of time and temperature makes up the **heat sum**, the number of days in the year (n) in which the mean temperature (Tm) exceeds the threshold temperature (Tt) for growth, multiplied by the mean temperature minus the threshold temperature. (The threshold temperature is about 0 °C for species in temperate vegetation but can be as high as 10 °C for tropical vegetation.) The result of the sum is called the **day-degree total** (D), when time is measured in days:

$$D = n\,(Tm - Tt)$$

From these three types of information we can build a 'decision tree'. Armed with meteorological records for a particular place, we can work through the 'tree' to reach a prediction of the type of natural vegetation at that place. A map drawn from these predictions agrees closely with published maps of observed distributions and with areas that were not included in the construction of the decision tree.

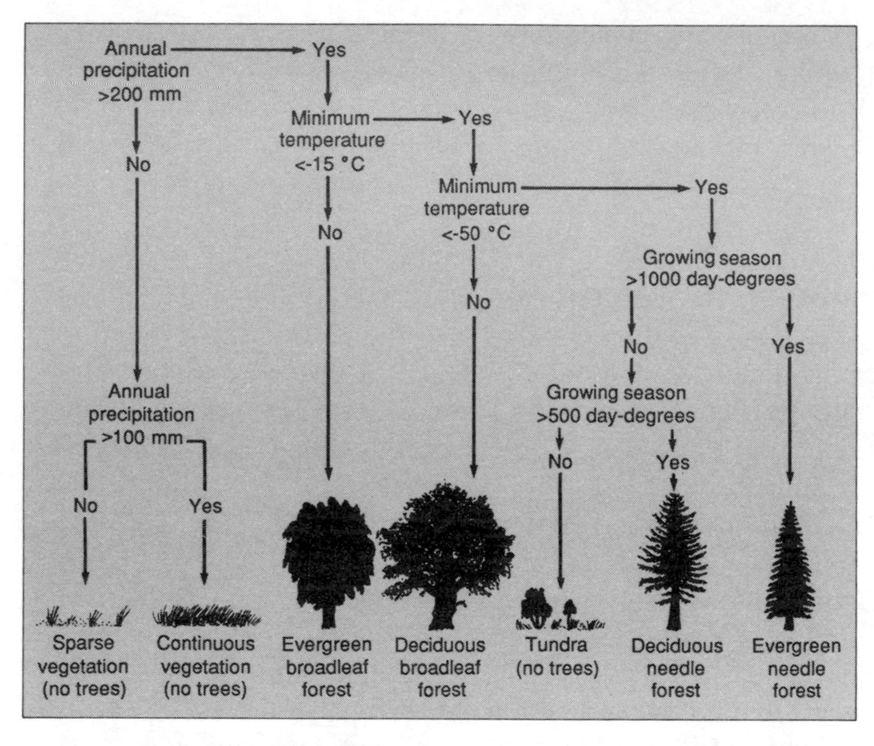

A decision tree allows you to predict the vegetation from information about the climate

The value of establishing the basic rules that govern how climate controls the type of vegetation is that they can be used to predict how vegetation will respond if the climate changes. One of the biggest changes climatologists expect in the next few decades is global warming – a symptom of the greenhouse effect. Around the equator the rise in temperature over the next 40 years or so will be small, maybe only 1 or 2 °C, but nearer the poles the increase will be much bigger.

The fauna of a region depends largely on the type and richness of the vegetation. If the vegetation changes, so will the fauna. Vegetation has another role in any particular ecosystem: it protects the soil by slowing or preventing erosion. Some types of vegetation, especially large areas of forest in humid climates, even control their own weather, increasing the amount of rainfall. Ecologists are understandably keen to predict how vegetation will change, because it has repercussions for all forms of life. But climatologists would also like to know what will happen to ecosystems because they interact with the atmosphere and can contribute to the greenhouse effect – or help to counteract it. The decision tree shows what the likely changes will be in plant forms in the greenhouse climate.

The greenhouse scenarios are figments of scientists' computer models. Several groups of researchers around the world have developed models to predict what will happen to global climate when the concentration of carbon dioxide in the atmosphere is double that of the pre-industrial era. Each model gives a slightly different picture – especially for the patterns of rainfall. But these simulations allow us to make some predictions.

By following the model designed by scientists at the Goddard Institute of Space Studies (GISS) in New York and using the decision tree, we can produce a map of how the world's vegetation might look in 50 years' time. Some areas will respond much more than others to the changes. The greatest change will be in the far north, where the vast, treeless tundra will shrink. Sandwiched between the ice cap and the great boreal forests, tundra grows where the soil is frozen for much of the year and the growing season is too short for trees. When the climate warms and the frozen soil melts, the coniferous forest will spread north, pushing the tundra up to the limits of the northern land mass. The loss of the tundra will in turn affect the animals that are adapted to survive its harsh conditions, or which migrate there in summer to breed, taking advantage of the brief flush of food. Many will probably not survive.

The warming of the tundra will unleash a great store of carbon that built up over thousands of years. Beneath the sedges, mosses and stunted shrubs lie the accumulated remains of dead plants in the form of peat. Peat accumulates because in the cold, wet conditions of the far north, dead leaves and plants decompose very slowly. As the soil warms and dries out, the process will become much faster, releasing carbon dioxide. The soils beneath the tundra contain some 14 per cent of the carbon stored in all the soils of the world. The sudden release

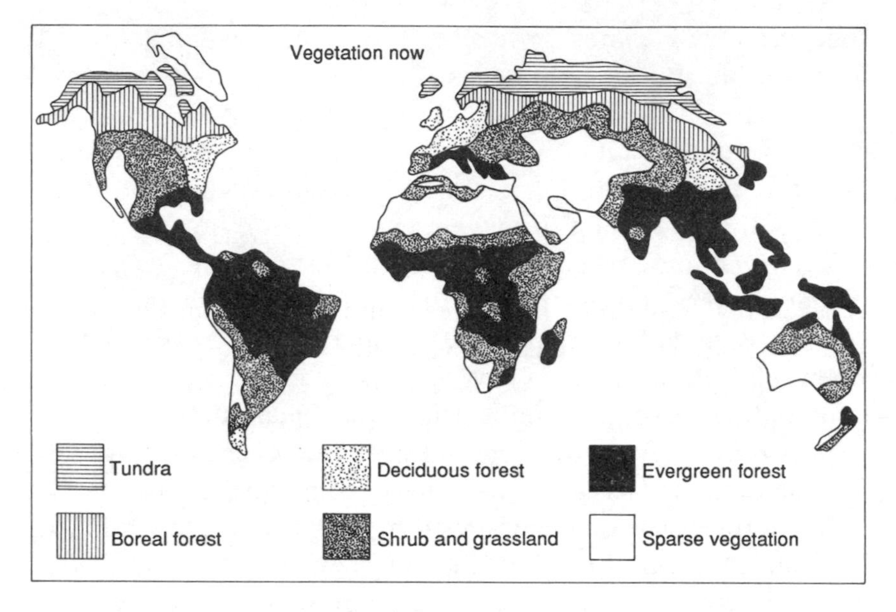

Today's world: a rough outline of potential natural vegetation types as we know them

Tomorrow's world: some types of vegetation will move, others will not

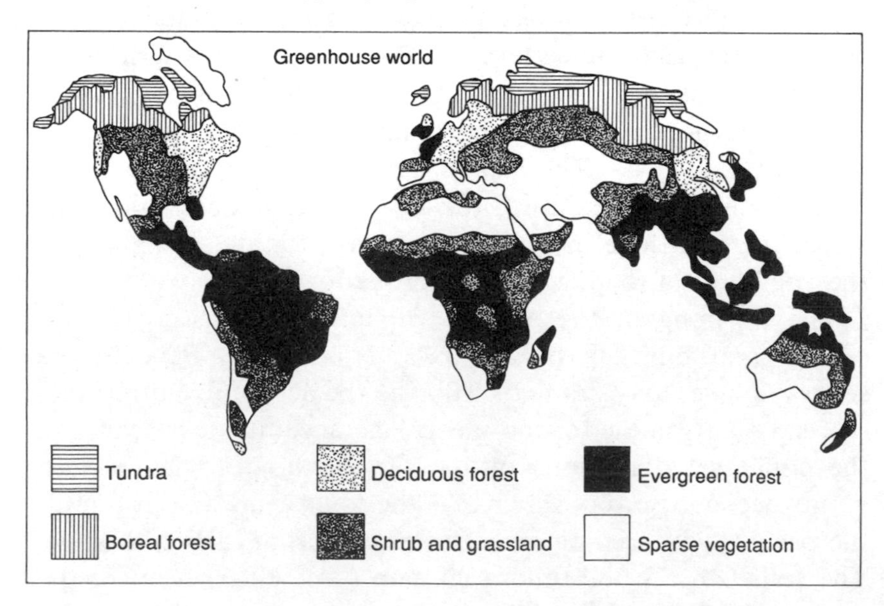

of carbon, as carbon dioxide, will increase still further the concentration of the gas in the atmosphere and contribute to the warming effect. Methane, another greenhouse gas, is also released from the tundra.

Another feedback mechanism will also hasten the loss of tundra: the surface of tundra vegetation reflects about 20 per cent of incoming solar radiation, while boreal conifers may reflect only half that. The forest, and the air above it, will warm as a result, aiding its spread northward.

Winter temperatures will increase more than summer temperatures. Many broad-leaved evergreen trees, such as the oaks of the Mediterranean region, will be able to survive the relatively mild winters of temperate latitudes – even as far north as Britain. Any spread of these species would be at the expense of broad-leaved deciduous species, the beeches, oaks and so on of western Europe. In North America, the evergreens of Florida and Texas will not push north; they are held in check by summer droughts and the rather small change in winter temperature.

The geographical range of the tropical and equatorial rainforests (based on temperature and rainfall in the wet season) will change little. Areas of southern Europe, eastern Asia, North and South America and South Africa, which botanists class as shrubland, will shrink. Sparse or desert-like vegetation will replace it.

Overall, the change of climate predicted by the GISS model will seriously damage some types of vegetation, particularly in the northern hemisphere. The change will be fast. The tundra will shrink noticeably within a decade. At the same time, crops in the American wheatbelt may fail as drought becomes a regular feature of the prairie climate. Elsewhere, changes might be for the better. The GISS model predicts more rain over Saudi Arabia and parts of Australia, prompting some scientists to speculate on a 'greening' of the deserts in these places.

Optimism in the air
The unseen fertilizer

The maps showing how vegetation will respond to a change of climate present a range of optimistic and pessimistic views of the future. In some areas the greenhouse effect will hardly be noticeable. Whatever happens to the climate, the increasing amount of one greenhouse gas, carbon dioxide, will have a more direct effect on plants. Carbon dioxide is one of the raw materials of photosynthesis. The more carbon dioxide in the atmosphere, the more productive the plants are likely to be. In effect, adding more carbon dioxide to the atmosphere is like giving the plant a fertilizer. This works in glasshouses, where growers of tomatoes and pot plants improve their yields by pumping carbon dioxide into the atmosphere.

Plants use one of two mechanisms of photosynthesis, and their response to an increased concentration of carbon dioxide depends on which type they use. Almost all trees and plants from cold climates, and most of the other species in the world, are what are known as **C3 plants**. The other plants are called **C4 plants**. This group is small but includes some important crops, such as maize and sugar cane. The C4 species have a specialized photosynthetic biochemistry which concentrates carbon dioxide within the leaf before converting it to carbohydrates. An increase in carbon dioxide outside the leaf will have little effect on the rate at which these plants photosynthesize. The other type of plant will benefit, and experiments show that if the two types are growing together, the C3 species flourish at the expense of the C4 species. In nature, C4 plants might die out in the face of the competition.

In most species of plant the pores in the leaves, the **stomata**, close as the concentration of carbon dioxide increases. Closing the aperture of the stomata reduces the rate of transpiration from the leaf, but because of the extra carbon dioxide it has little effect on the rate of photosynthesis. Plants that lose less water while photosynthesizing at the same rate are more efficient, so they benefit in this respect. There are exceptions.

Some conifers and some tropical species show virtually no stomatal response to extra carbon dioxide. Although they will increase their rate of photosynthesis, they will not increase their efficiency as much as plants that are sensitive to carbon dioxide.

Biologists face a big problem when they try to predict what will happen to plants based on the results of experiments in glasshouses or in a small corner of an experimental plot. The environment in a glasshouse, or any other controlled environment, is vastly different from a natural environment. There is also a problem of scale. Experiments on a few plants grossly oversimplify what will happen within complex vegetation. The plants themselves influence the microclimate and the behaviour of other plants. The stature of the vegetation and the smoothness of its surface determine how much local weather patterns control the behaviour of the plants, for example in how much they transpire. In forests, which have a rough surface, the local weather mixes with the microclimate inside the forest and so controls the rate at which the forest transpires. In contrast, a grassland is aerodynamically 'smooth', and there is little mixing of the local weather above it and the microclimate within it. As a result, the grassland microclimate is more or less independent of the local weather.

Because we still know very little about these interactions it is impossible to say how entire tracts of vegetation will respond to the increase in atmospheric carbon dioxide. Yet, there are some hints of what is to come. Leaving the models aside, some scientists are looking directly at the annual increases in the concentration of carbon dioxide – and the responses vegetation is showing. This technique is based simply on an analysis of the annual trend in the partial pressure of carbon dioxide at particular sites around the world. Charles Keeling, of the Scripps Institute of Oceanography in California, established the first monitoring station for carbon dioxide in Hawaii in 1958. The records from Hawaii show without a doubt that the greenhouse effect is real.

If extra carbon dioxide does have any effect on the rate at which plants photosynthesize, it ought to show most clearly in vegetation with markedly seasonal activity. In tundra, boreal forests and deciduous forests, for example, there is a period during winter when plants do not photosynthesize and so do not take up any carbon dioxide; photosynthesis resumes in summer when the plants take up significant amounts of carbon dioxide. Measurements of the amount of carbon dioxide in the air near such vegetation should show significant variations between summer and winter. In contrast, tropical and equatorial forests, which grow all year, will show only small seasonal fluctuations that depend on the amount of rain. Measurements taken at Point Barrow, in the remote Alaskan tundra, show that in the past decade the amount of carbon dioxide has increased significantly every year. The figures also show that the amount of carbon dioxide used in photosynthesis between June and November, the **drawdown**, has also increased. This suggests that the vegetation around Point Barrow is photosynthesizing more in response to the increasing concentration of carbon dioxide in the atmosphere. These observations contradict the results of glasshouse experiments, which show that the tundra could benefit only a small amount from the fertilizing effect of carbon dioxide, because tundra suffers a marked shortage of the nutrients a plant needs in order to grow.

The plants of the tundra may be able to make use of extra supplies of carbon dioxide. Yet this will not protect it. The coniferous trees of the boreal forest will also benefit from the extra carbon dioxide. Because they grow much taller than the dwarf plants of the far north, they will intercept more sunlight and thrive at the expense of the tundra.

6 May 1989

Further Reading

Climate and Plant Distribution by F. I. Woodward (Cambridge University Press, 1987) provides details of how vegetation responds to climate. The definitive study of the greenhouse effect is *The Greenhouse Effect, Climatic Change and Ecosystems* edited by Bert Bolin, Bo Döös, Jill Jäger and Richard Warrick (Wiley, 1986).

PART 3

Life and Cell Biology

CHAPTER 17

Life and the Universe

Nigel Henbest and Heather Couper

In November 1974 scientists in the US broadcast a radio message. Its content was not very exciting: descriptions of the materials that make up living cells and of the human population of the Earth. What was unusual was its destination. The scientists did not send their broadcast in the direction of any listener on the Earth. They sent it towards a cluster of stars some 24 000 **light years** away – in the hope that some alien being might tune in. (A light year is the distance that light travels in a year, equal to 9.4605 million million kilometres.)

Why were the scientists so confident that there are other intelligent beings elsewhere in the Universe? After all, we have not yet met any extraterrestrial beings – except in science fiction films and television series! The answer lies in what we know about life on the Earth, and what astronomers have found out about the Universe in general. We have learnt that the basic materials of life as we know it (carbon compounds) are common in the Universe – not just on planets, but in comets and even floating freely in giant clouds in space. We have also found that the Sun is a very ordinary star, in terms of its size, its mass and its temperature. There are literally billions of stars like the Sun in our Galaxy, so there is a good chance that some will have planets like the Earth, where life can evolve.

Philosophers have argued for centuries about how to define 'life'. But if we study living things on the Earth, we find that there are some very basic characteristics that will guide us when we consider life elsewhere in the Universe. Life on Earth shows an amazing diversity in its appearance, from whales to widow spiders, from sea slugs to sequoias. Yet all these incredibly different forms of life share a common building block. They are all made of the same kind of basic units: cells. Each cell is an intricate structure, made up of many complex molecules – for example, the DNA molecule that contains the blueprints of the cell and passes them on to subsequent generations. But these molecules all have one common characteristic. They are built up from long chains of atoms of one particular element: **carbon**. (Some writers of science fiction have suggested that alien life forms could be based on a different element, for example silicon. However, no other elements seem to be able to build up into the highly complex compounds that are required to carry out all the functions of a living object. Certainly, that is true for the forms of life that we know of.)

Carbon is the fourth most abundant element in the Universe. Astronomers believe that the cloud of gas that gave birth to the Sun and the planets was rich in carbon and its compounds. On the early Earth these simple compounds grew into larger molecules like DNA, and these came together to make the first

Life on planet Earth evolved from simple chemical compounds to complex organisms. The Earth's original atmosphere contained gases such as carbon dioxide, water and ammonia (1), erupted from volcanoes. Strokes of lightning welded these into the raw materials of life, such as amino acids (2). In small pools, these turned into long chains, proteins and DNA (3), which formed into cells (4) – the first living organisms. Through evolution, cells joined up to form more and more complex organisms, which changed the atmosphere to its current composition of oxygen and nitrogen (5). The end point of evolution, so far, is human beings (right)

cells. From then on, life evolved along a number of bewildering routes to form the wide diversity of plants and animals that we see around us today. Eventually, one species developed a special mental skill: intelligence. It has allowed human beings to look around the Universe and to attempt to determine whether life has developed on planets other than the Earth.

The beginning of life

No one can really tell how life began on the Earth, soon after its formation some 4600 million years ago. But there are enough clues to give us a rough idea.

Scientists have long known that the 'raw materials' for life on Earth are compounds of carbon. In the past 20 years radio astronomers have picked up characteristic radiations from more than 60 types of molecule containing carbon in space, mainly from dense dark clouds where stars and planets are being born. Astronomers have also found such chemicals in meteorites, which are fragments of

Compounds of carbon, the building blocks of life, form quite readily in the Universe. Even in the space between the stars, astronomers find molecules containing as many as 11 carbon atoms (A). When planets like the Earth are born, they are surrounded by atmospheres consisting of quite simple compounds (B), but the action of lightning, ultraviolet radiation and the shock of meteorite falls easily change these into more complex material (C)

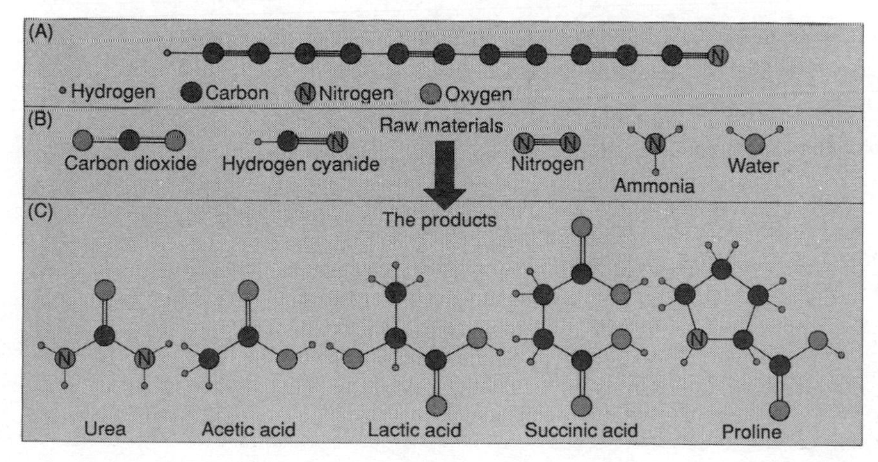

asteroids that never built up into a planet, and in Halley's Comet, a chunk of ice from the outer part of the Solar System.

Meteorites and comets could have coated the early Earth with a rich layer of carbon compounds. In addition, lightning could have welded together simple molecules in the atmosphere to make more complex chemicals. In the early 1950s two American scientists, Stanley Miller and Harold Urey, replicated this in the laboratory: they passed electric sparks through a mixture of gases and ended up with a cocktail of carbon compounds.

When the Voyager 1 spacecraft flew past Saturn in 1980, it found that Saturn's largest moon, Titan, is covered with a haze of orange cloud, probably made of carbon compounds. Astronomers believe these are similar to the chemicals formed in the early atmosphere of the Earth – so Titan is 'an Earth in deep freeze' – but conditions were so cold that these compounds never built up into living cells. An American–European space mission, called Cassini–Huygens, will drop in on Titan early next century to find what the clouds are made of.

Most scientists believe that these simple compounds dissolved in the waters of the early Earth. In ponds by the seashores, this 'primeval soup' became more concentrated, and reactions turned these compounds into complex chemicals that eventually formed the first living cells. These cells evolved into more complex organisms.

The place to start such a search is our Solar System. As well as the Earth there are eight other planets circling the Sun. In past centuries many astronomers believed that all the planets had some sort of life. Sir William Herschel, who discovered Uranus, even believed there might be life on the Sun! But the other planets – except one – have turned out to be either far too hot or much too cold. The only planet that holds out any hope for life is Mars. Science fiction writers have loved to terrify their readers with stories of Martians invading the Earth. A hundred years ago astronomers such as Percival Lowell in the US and Giovanni Schiaparelli in Italy thought they could see artificial 'canals' on Mars, and Lowell built a huge observatory in Arizona in 1894 to see them better. But in the 1960s the Americans sent the first in a series of Mariner spacecraft to Mars. Their close-up pictures showed that the canals were only an optical illusion. In fact, Mars is a barren and frozen world, with only a very thin atmosphere. Later spacecraft, however, revealed a more interesting side to Mars: mountains, canyons and volcanoes, all larger than their counterparts on Earth. Most exciting, there were valleys that looked as if they had once contained rivers. This showed that Mars was once warmer, with liquid water on its surface. In that case, life may have begun there – and some living cells might even now be preserved in its frozen soil.

So American scientists built a pair of very sophisticated unmanned spacecraft: the Vikings. Launched amid much optimism in 1975, the Vikings landed on Mars the following year. Each carried a small laboratory that warmed up a sample of soil, and tested it for signs of life. The results were negative.

Too hot or too cold
Barren neighbours

Whether life ever did develop on Mars or not, the planet has definitely suffered a planetwide ice age that may never thaw. Our other next-door neighbour, Venus, has gone the other way. There are signs that it too once had seas of liquid water, but it is now searingly hot, with a temperature of 470 °C, which varies little from its poles to its equator. It provides us with an example of the **greenhouse effect** at work. Carbon dioxide gas trapped the Sun's heat; this released more carbon dioxide from Venus's rocks, and the temperature rose until the oceans boiled away. The Earth has suffered neither the fate of Mars nor that of Venus, because it is just the right distance from the Sun. If it were a fraction closer, life would have been burned up by the greenhouse effect that has happened on Venus. Just a mite further away, and we, too, would have suffered the fate of frozen Mars.

But distance from the Sun is not everything. After all, the Moon is at the same distance from the Sun as the Earth, but it has no life. The problem is that the Moon is small, so it has only a weak gravity that cannot hold on to an atmosphere. As

Spacecraft from another world?

On 30 December 1978 an Argosy freight plane took off from Wellington, New Zealand. During this routine flight the captain spotted something he had never seen in 23 years of flying – an unidentified flying object (UFO). By chance, an Australian TV crew was on board the plane and managed to film it.

Spurred on by well-publicized cases like this, groups of watchers gather every night throughout the world to keep a vigil for UFOs. These people are convinced that they are seeing 'flying saucers' – craft from other worlds.

The modern rash of UFO sightings began 40 years ago, when an American pilot reported an airborne object as looking like a 'flying saucer'. Since then, there have been thousands of sightings the world over, with hundreds of photographs of supposed alien spacecraft. Some people claim to have had more bizarre experiences, in which they have been abducted by the inhabitants of these craft – and in some cases subjected to horrific experiments.

Most scientists, however, dismiss these claims. Astronomers find that most people who see UFOs are not used to looking up at the sky, and they misinterpret what they see. To back this claim they point out that backyard stargazers, who are forever looking up at the sky, never see UFOs.

When investigators have looked in detail into sightings of UFOs they have generally found quite logical explanations. Some are just frauds. Many more are simply sightings of natural or man-made objects that people do not recognize. A weather balloon, catching the rays of the setting Sun, can easily be classed as an inexplicable light in the sky. The 'UFO' seen by American President Jimmy Carter was the planet Venus, and the New Zealand UFO turned out to be the planet Jupiter.

a result, there are no gases for any life as we know it to breathe; and it also means that any water on the Moon would instantly boil away into space. The depressing conclusion is that there is no life elsewhere in the Solar System. However, our Sun is only one of 200 000 million stars in our Galaxy, so there is a good chance that life may have developed on a planet like the Earth which is going around another star.

Astronomers have been keenly interested to find out whether other stars actually have planets. The problem is that planets are small and dark, and lie close to their parent stars, which are big and bright. It is like looking for dark moths flying around a distant streetlight. No telescope yet built can show a planet like Earth circling another star. But astronomers are resourceful. At observatories the world over they have found indirect ways to flush out planets skulking around nearby stars. The work requires patience because the astronomers are searching for the effects caused by a planet over a whole orbit, which may take years. Some astronomers investigate a star's path through space. If the star is 'wobbling' from side to side as it moves, it must be feeling the pull of an unseen planet that is orbiting the star. Other astronomers have looked for the effect that such a wobble has on the star's speed towards or away from us (as measured by the Doppler effect, the observed change in wavelength of the radiation emitted by a star because of its motion). In 1988 a team of Canadian astronomers, led by Bruce Campbell, claimed that half the stars they had been studying showed signs of a wobble, indicating that they had planets. Another way is to look at very young stars, searching for signs of a planetary system being born. Astronomers now think that planets were born along with their parent stars from a disc of gas and dust that surrounded the star. It should be easier to see this big disc than to detect the compact planets that eventually form within it (see Chapter 6).

In 1983 astronomers launched the first satellite to carry a telescope to pick up infrared radiation from space. The Infrared Astronomical Satellite could 'see' much fainter sources of radiation than any telescope on the Earth. To the astronomers' surprise it detected infrared radiation coming from the vicinity of several nearby stars, including Vega, the fifth brightest star in the sky. The spectrum of the radiation showed that it was coming from matter much cooler than the star's surface, probably in the form of a disc of small solid particles (**dust**) encircling the star. Astronomers in South America soon confirmed this hypothesis when they photographed the faint disc surrounding the star Beta Pictoris. These discs seem to be planetary systems in formation. Once again, this indirect evidence suggests that planets may be quite common in the Universe.

Anyone out there?
Search for radio signals

It is a long step, however, from finding other planetary systems to being sure that there is intelligent life 'out there'. The system must contain a planet of the right size at the correct distance from the star; the compounds on its surface must react in such a way as to form the chemicals of life; the cells must evolve to produce complex beings, which must acquire intelligence. Many scientists have attempted to estimate the probability of all these things happening – and the calculated number of intelligent civilizations ranges from 10 million down to only one!

If there are many intelligent civilizations in the Universe, then they should be in contact with one another. Following this line of thought an American radio astronomer, Frank Drake at Green Bank in West Virginia, set up Project OZMA in 1960. This was the first search for artificial radio signals from space. Since then, radio astronomers around the world have been scanning the skies on the lookout for signals. Unlike light, radio waves can travel through the murky corners of our Galaxy without being absorbed by dust – as far as radio waves are concerned, space is virtually transparent. So, most scientists assume that if another advanced civilization elsewhere in the Universe were trying to signal its existence, 'it' would broadcast

Space phone book

In the 1960s the American radio astronomer Frank Drake proposed a simple method of calculating the number of alien civilizations that are currently trying to contact the Earth. He summed it up in an equation:

$$N = R \times f_p \times n_e \times f_l \times f_i \times f_c \times L$$

We are trying to find N, the number of communicating civilizations that exist now, by putting into the formula figures for: R, the rate at which stars are being formed; f_p, the fraction of stars with planets; n_e, the average number of planets in the system that are suitable for life; f_l, the fraction of these planets on which life develops; f_i, the fraction of the life-bearing planets that develop intelligent life forms; f_c, the fraction of these planets with intelligent life on which the capacity for interstellar communication develops; L, the average lifetime of such a 'technological civilization'.

Astronomers agree that R is about 10 new stars per year: but there is no agreement at all on the other factors!

An optimist could say that all stars have planets ($f_p = 1$), with one that is suitable for life ($n_e = 1$); and that life is sure to develop ($f_l = 1$) and evolve to produce intelligent life ($f_i = 1$) that wants to communicate with others ($f_c = 1$). If the lifetime of the civilization is a million years ($L = 1\,000\,000$), then we can expect 10 million civilizations to be broadcasting messages through the cosmos.

On the other hand, a pessimist could put in very low figures. We could argue that only one star in ten has planets ($f_p = 0.1$); that in each planetary system there is only a one-in-ten chance of finding a suitable planet for life ($n_e = 0.1$); that chances are low of life occurring (say $f_l = 0.1$), developing intelligence ($f_i = 0.1$), wanting to communicate ($f_c = 0.1$) and escaping the global war that can come with advanced technology (perhaps $L = 100$ years). Then $N = 0.01$. In other words, there is only a slim chance of a communicating civilization evolving at all in our Galaxy, and we are likely to be the only one at present.

its messages at radio wavelengths. Out of 55 searches that astronomers have made, 48 were for radio waves.

Over the years there have been some heart-stopping moments. One was the occasion when a radio telescope in Ohio picked up a very strange signal. A euphoric astronomer, convinced he had picked up an extraterrestrial message, scrawled 'WOW!' across the paper chart. He looked the next day, in the same part of the sky. And the day after that, and the day after that. It never came back. Then there was the discovery of rapid pulses of radio waves coming from an object hundreds of light years away. Jocelyn Bell and Antony Hewish, researchers at the University of Cambridge, were not looking for intelligent signals from space in 1968, and they jokingly called the signals 'LGM-1' – for 'Little Green Man'. Later, it turned out that the 'signals' were entirely natural, and came from rapidly spinning neutron stars (the compressed cores of stars that have exploded as supernovae, see Chapter 6).

Since Project OZMA our technology has improved dramatically. New types of electronics and computer make the search for signals much easier and quicker. For example, we do not know what frequency the aliens might be broadcasting on. This was a problem for Drake, who had to make a guess. Now several teams in the US are building multichannel analysers that can tune into millions of extraterrestrial 'radio stations' at once. Yet with all this sophistication and dedication, the fact remains: we have not picked up any known alien signals. Some scientists are not deterred, however. They point out that these searches involve only 'listening in'. But if *everyone* has been 'just listening', then there will be no messages for anyone to pick up.

That is why a group of American scientists sent a message into space in 1974. They used the world's biggest radio telescope – a 'dish' 300 metres across, near Arecibo in Puerto Rico – to beam the message to a cluster of stars in the constellation Hercules. But the message will take 24 000 years to get there, so even if someone picks it up and sends a reply, we will not receive it until the year AD 50 000!

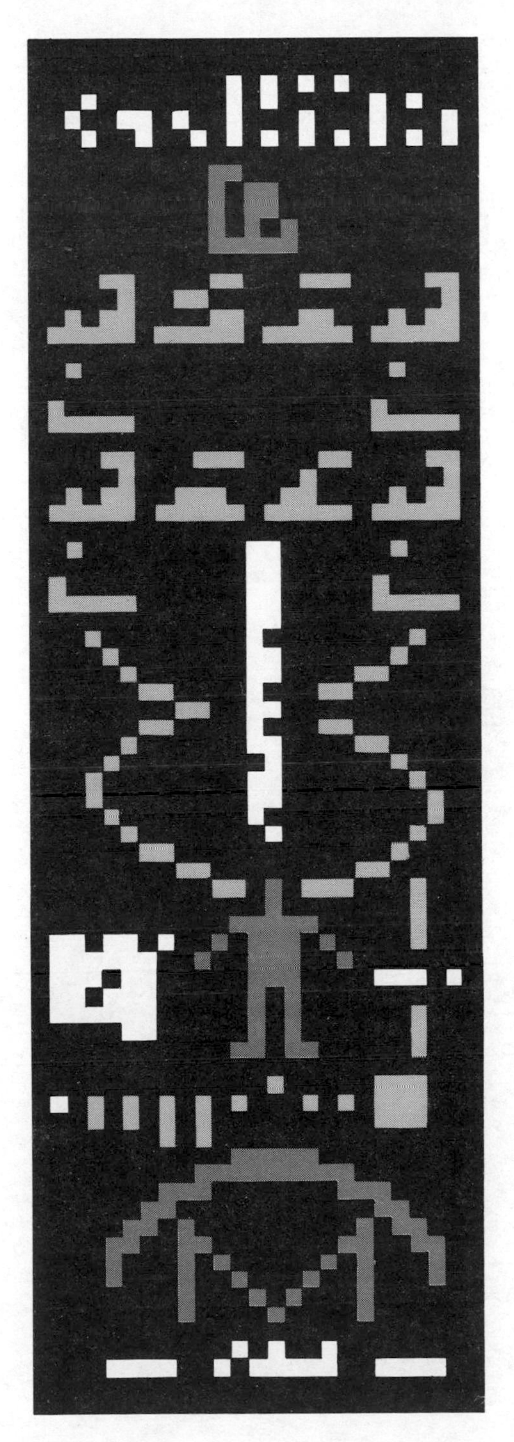

Say Hello! A radio message to the Cosmos: in 1974 astronomers broadcast a string of radio 'pulses', which made up the pattern shown here. The first line shows numbers from one to ten in the 'binary' system that computers use. The next three lines indicate the chemical compounds that are important in living cells, while the double spiral below depicts the shape of the DNA molecule that is the 'library' of a cell. The human figure is flanked by binary numbers showing the height of a human (right) and the population of Earth (left). The next line indicates the planets of the Solar System, with the Sun at the right and the Earth displaced upwards to draw attention to it. The bottom part depicts the radio telescope that sent the message into space

Probing deep space
Revealing our presence

Scientists have also sent messages into space on the side of four spaceprobes that will ultimately leave the Solar System. The Pioneer 10 and 11 spacecraft each carry a plaque advertising our whereabouts in the Galaxy. Voyager 1 and 2 each have a long-playing record that plays the 'Sounds of Earth' – including greetings in many languages, with pictures also encoded into its grooves.

Some scientists have said we should not be trying to contact other civilizations. If we succeed, and 'they' are much more

This plaque was affixed to the Pioneer 10 and 11 spaceprobes, which are leaving the Solar System. The left half is a map showing where the Sun is to be found. All the sizes are expressed relative to the wavelength of 21 centimetres emitted by a hydrogen atom, shown at top left

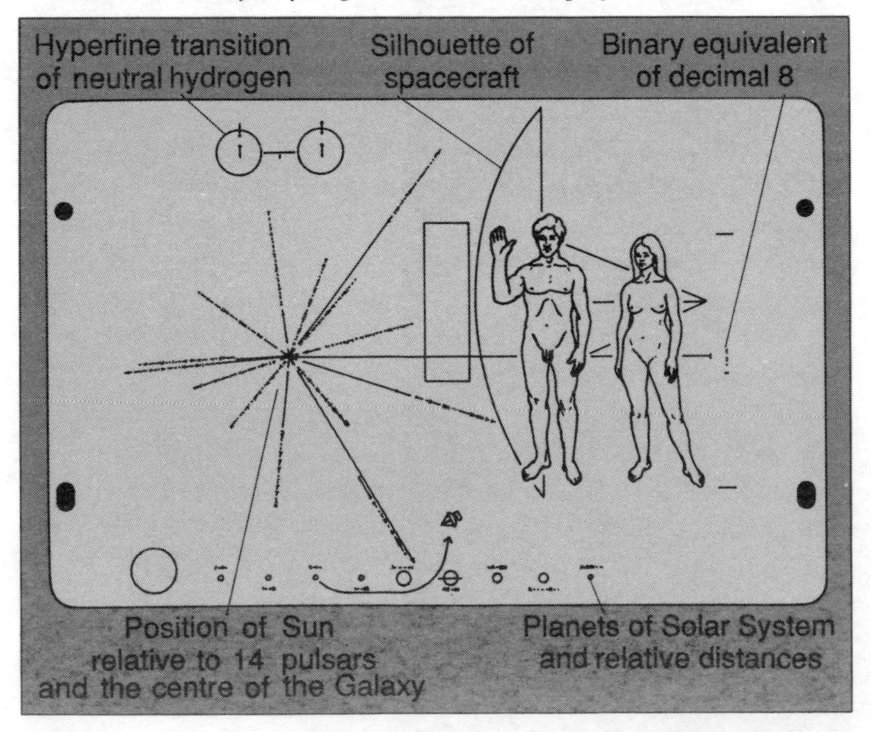

advanced than us, then our culture might be overwhelmed – just as the native cultures of South America or the Pacific have disappeared under the onslaught of European culture. But it is already too late. We have been broadcasting our presence unintentionally for the past 60 years. First radio, and now the far more powerful television signals, are winging away from Earth at the speed of light. Anyone on a planet orbiting a star within 60 light years of the Sun could know of our existence. So, perhaps we can expect a message soon – even if it's just fan mail for *EastEnders*, *Dynasty* or *Neighbours*!

18 March 1989

Further Reading

Life off Earth by Ian Ridpath (Granada, 1983) and *The Search for Extraterrestrial Intelligence* by Thomas McDonough (Wiley, 1987) provide simple introductions to the search for life in space. *The Quest for Extraterrestrial Life* edited by Donald Goldsmith (University Science Books, distributed in the UK by Oxford University Press, 1982) is a more advanced text – about A-level – and contains many important, original articles.

CHAPTER 18

The Human Immune System: Origins

Linda Gamlin

The longest that anyone has lived without an effective immune system is 12 years. Doctors at a hospital in Houston, Texas, managed to keep a boy born with 'severe combined immunodeficiency' (SCID) alive for this long. But to do so they had to enclose him in the microbe-free world of a plastic bubble soon after birth. The boy had no direct contact with anyone. Nurses passed sterilized food in through an airlock, and he breathed filtered air to ensure that no bacteria, fungi or viruses reached his lungs. Doctors hoped to cure the boy eventually by transplanting bone marrow from a relative – bone marrow is the source of all the cells that make up the immune system. But the operation was unsuccessful, and soon afterwards the boy died.

The plight of 'the boy in the bubble' illustrates how much we need an immune system. To most microorganisms our bodies are a huge potential source of food, warmth and living space. Bacteria, viruses, fungi and other microbes are constantly trying to invade us, and to remain healthy we must fight off millions of them daily. The familiar disease-causing organisms or **pathogens** – the viruses that cause influenza or measles, for example – are just the tip of the iceberg. These are the 'professional' invaders, the ones we are less successful against. It takes time to combat them, so they cause a disease in the process. There are many more microorganisms, in water, soil, air and food – even in our own digestive tract – that do not cause any disease in normal, healthy people. But when the immune system is defective – as in SCID,

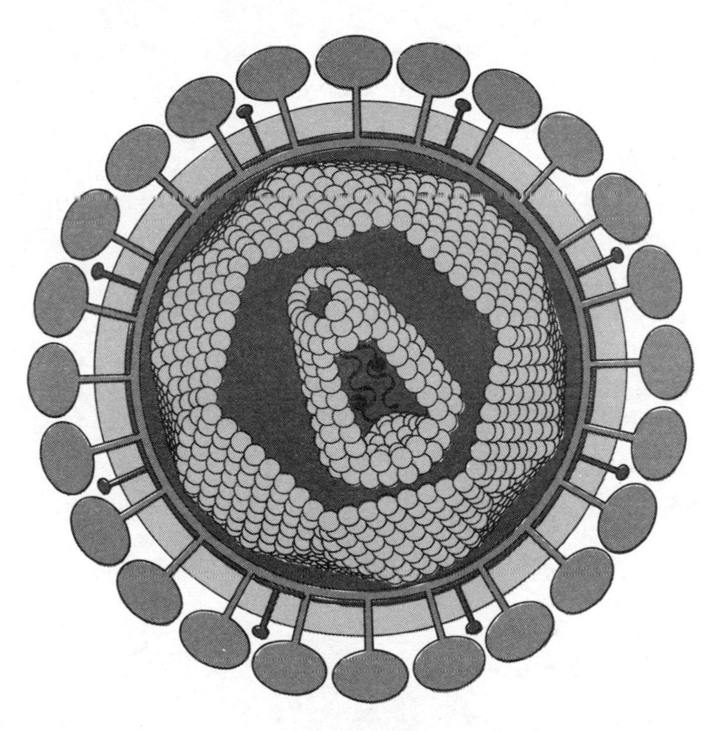

Profile of a virus – the latest model of HIV; it attacks the immune system itself, causing AIDS

or in AIDS – these organisms can invade the body, producing what are called 'opportunistic infections'. Most children with SCID die of such infections before they are a year old.

Barriers to infection
The front line

Faced with this onslaught of microbes, how does the normal human body defend itself and stay healthy? To begin with, it keeps out as many potential pathogens as possible with barriers such as the **skin**, and other non-specific defences. The skin, which is waterproof, is impenetrable to most invaders, and it produces fatty acids that many microorganisms find toxic. Areas not covered by skin, such as the eyes, mouth, lungs and digestive tract, are more vulnerable, but they have alternative defences. Tears, saliva, urine and other body secretions contain **lysozyme**, an enzyme that can kill certain types of bacteria by splitting the molecules found in their cell walls. Mucus in the nose and airways engulfs bacteria and stops them penetrating the membranes. Cilia – tiny beating 'hairs' – then push the mucus out of the airways into the throat, where it is swallowed. In the stomach, acid kills most of the microorganisms in food, as well as starting the process of digestion.

Helping us to keep pathogens out are a vast population of harmless **commensal bacteria**. The bulk of these live in our intestines, where they benefit from free food. These bacteria – the 'gut flora' – unwittingly help to exclude more harmful microbes by filling all the available ecological niches in the gut.

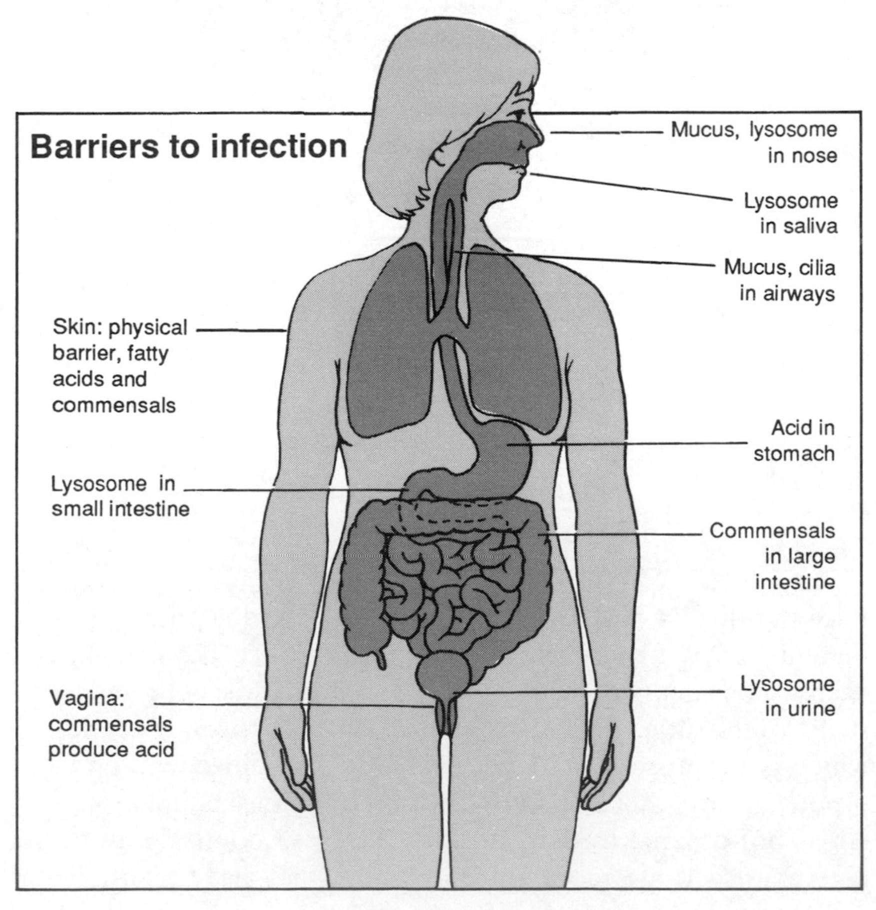

Other commensals live in the vagina and on the skin. The vagina secretes a carbohydrate which the bacteria feed on, producing lactic acid in the process. This makes vaginal secretions acidic and thus hostile to many fungi, bacteria and viruses.

Microorganisms that breach these outer defences are much more difficult to deal with, because the body must somehow distinguish them from its own cells. It also has to kill them without doing too much damage to its own tissues – a task that is far from easy.

Evolution
Layers of complexity

Human beings are not alone in facing these problems. All animals, even the very simplest sponges and worms, have to defend themselves against attack. By looking at the immune systems of such animals we can see that some parts of our immune defences resemble theirs. This tells us that our immune system has evolved very gradually, over hundreds of millions of years, from simple beginnings to its present complexity. At the same time, certain microorganisms have been continually evolving new ways to overcome our defences. The 'biological arms race' with these pathogens has done much to shape our immune system.

During the evolution of the immune system, new types of immune cells have emerged to add to the original, comparatively simple system of lower animals. New control systems have developed to keep these cells in check, and the new and old systems have become integrated so that they work together. This has produced an immune system of quite extraordinary complexity in vertebrates, particularly mammals. Because the system has built up bit by bit, it often seems unnecessarily complex and rather illogical. But the important thing is that it works – most of the time.

Among the more recent evolutionary additions are the

molecules called **antibodies** – the one part of the immune system that most people have heard of. Special cells known as B cells, or B lymphocytes, produce the antibodies. The family of cells to which they belong, the lymphocytes, are relative newcomers to the immune system. They have been around for a mere 400 million years, since the early vertebrates appeared on Earth.

The most 'primitive' type of immune cell is one that engulfs and digests invaders in the same way that an amoeba obtains its food – by phagocytosis. All invertebrates, from sponges upwards, have immune cells of this type, called **phagocytes**. Once they have engulfed invading microorganisms, phagocytes usually kill them, although some cunning types of bacteria block the killing mechanism and thrive inside the phagocytes. There are two methods of killing the invaders: by digestive enzymes, or by chemical reactions, controlled by enzymes, that release toxic products. Small packages within the cell, known as **lysosomes**, contain the death-dealing enzymes. These lysosomes give the phagocytes a granular appearance.

There are two main types of phagocyte. The first consists of large cells with a single, horseshoe-shaped nucleus – the **monocytes** and **macrophages** – the second of smaller cells with an irregular, many-lobed nucleus, known as **polymorphonuclear neutrophils**, or **PMNs**. The two groups have slightly different roles within the body. Macrophages are the principal 'rubbish collectors' in the body's tissues. They develop from monocytes, which occur in the blood but make up only 6 per cent of the white blood cells, or leukocytes (the cells in the blood responsible for immunity). The main function of these monocytes is to replenish the macrophage population. After leaving the bone marrow a monocyte circulates for one or two days in the blood, before squeezing between the cells of the blood-vessel wall and migrating into the tissues. There it develops into a macrophage.

Macrophages may continue to wander through the tissues, eating up microorganisms and other foreign bodies that they find. (Those in the lungs, for instance, engulf dust and fibres as well as microorganisms.) Other macrophages settle down and attach themselves to certain tissues, principally in the liver, kidney and spleen. These cells lie in wait: they are ready to engulf any invaders or debris travelling in the body fluids that flow past them. Known as the **reticuloendothelial system**, these sedentary macrophages play a vital role in keeping the body clean internally.

Polymorphonuclear neutrophils, the second type of phagocyte, are much more common in the blood than monocytes, making up 60 per cent of all white blood cells. Every minute of the day the bone marrow produces 80 million of these cells. They, too, migrate through the walls of the blood vessels into the tissues and are an essential part of our immune system, wiping out many infections before they have a chance to get going. Unfortunately, researchers use several different names for these important cells, including neutrophils, polys, neutrophil polymorphs, and granulocytes.

PMNs, unlike macrophages, are very short-lived, surviving for no more than a few days. They move towards sites of infection attracted by various chemicals, including some bacterial products and substances that escape from our own cells when the cells are damaged. The PMNs arrive first, but at a major site of infection the long-lived macrophages take over from the PMNs later.

Complement proteins
Helping the phagocytes

Phagocytes are an excellent form of defence, particularly against bacteria. But to be useful to the body they must 'recognize' their target. If they engulfed cells at random, they would cause enormous damage to the body's own cells. So how do phagocytes identify the enemy? The cell walls of bacteria are very varied, but there are some chemical features common to all, and others that are characteristic of particular groups of bacteria. Several of these chemical markers stimulate phagocytes

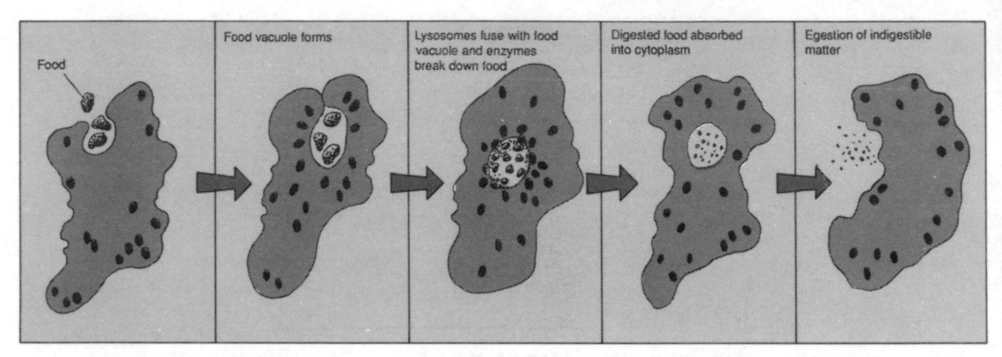

The process of phagocytosis: a giant amoeba engulfs and digests particles of food – the phagocytes of the immune system work in the same way

into action. Exactly how they do this is unknown, but the inherent 'stickiness' of bacteria is probably involved. Bacteria have a habit of sticking to the outer membranes of other cells – a necessary prelude to invading the body. By sticking to a phagocyte, bacteria may unwittingly stimulate it to engulf and digest them. Phagocytes that can respond to the chemical markers of bacterial cell walls probably represent the simplest type of immune system – some distant invertebrate ancestor of ours could well have developed such cells for its defence. Although the rest of our immune system is the product of hundreds of millions of years of evolution, large parts of it are just a refinement of this system. One of the main roles of antibodies, for example, is to expand the range of invaders that phagocytes can recognize, and improve their ability to engulf those invaders.

One of the earliest refinements to this phagocytic defence force may well have been a group of proteins called **complement**. These proteins occur in the blood and participate in immune reactions. Macrophages and monocytes synthesize complement proteins, but most come from cells elsewhere in the body, particularly the liver. The complement proteins are normally inactive, but in the right circumstances they interact to produce various defensive proteins. Some of these complement products stick to invading microorganisms, particularly bacteria and yeasts, making them recognizable to phagocytes as aliens.

How the complement proteins work

The central complement reaction is the splitting of a protein, known as C3. Various enzymes can split this molecule, and the reaction occurs spontaneously all the time.

Splitting exposes a reactive surface on C3b which can stick to a bacterial membrane or yeast cell wall. Another protein, called B, can bind to C3b, and when split by one of the proteases forms the new protein C3b,Bb. This acts as an enzyme, and splits C3 very efficiently so it amplifies the original response. But C3b,Bb is quickly inactivated by another protein unless it is bound to the right sort of membrane – so complement is unleashed only when needed.

As the reaction proceeds, the microbe acquires a coat of C3b, and C3b,Bb. Phagocytes have receptors for C3b, so they bind to it and engulf the invader.

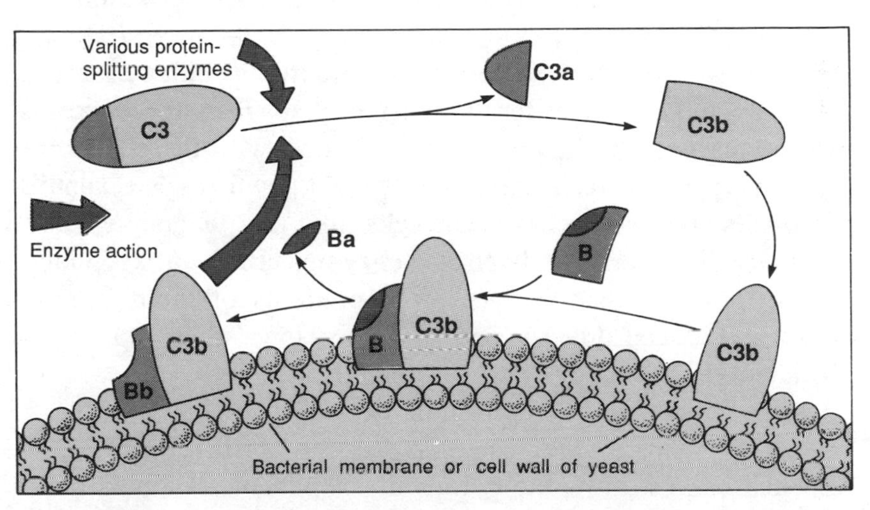

Others are powerful enzymes, which form a 'membrane attack complex' and destroy the membranes of invading microorganisms. A third group of protein fragments attract phagocytes and stimulate them to become more active.

Microorganisms activate the complement system in two distinct ways, known as the classical pathway and the alternative pathway. To trigger the classical pathway, they must first bind to antibodies. The alternative pathway probably evolved much earlier and allows certain microorganisms to activate complement directly, without the involvement of antibodies. The names of the complement pathways are misleading – they merely reflect the order in which immunologists discovered them. From an evolutionary perspective the alternative pathway is much older. The classical pathway is a new set of reactions, grafted on to the original set to allow complement and antibodies to interact.

Roots of allergy
Killing larger parasites

Bacteria, viruses and yeasts are much smaller than mammalian cells, so our phagocytes can easily engulf them. But not all invaders are this small. There are many multicellular parasites, such as flukes (parasitic flatworms) and nematode worms, whose size defeats the phagocytes. To deal with these parasites special types of cell have evolved, known as **eosinophils**, **basophils** and **mast cells**. These cells, like the phagocytes, have granules that contain defensive enzymes. But, unlike phagocytes, they can readily release the contents of their granules outside the cell (**degranulate**). This allows them to attack invaders that phagocytes cannot engulf.

With **eosinophils** the main defensive agent in the granules is an enzyme called major basic protein, or MBP. This powerful enzyme breaks down the body wall of parasites such as blood flukes. The eosinophil has protein receptors that bind specifically to one of the complement proteins. The complement protein concerned binds to these parasites, so the receptors allow the eosinophil to bind to its target via the complement protein. Eosinophils also have receptors for antibodies, which act in the same way as the complement proteins – as 'middle men' binding to the parasite. Once the eosinophil has bound to its 'middle men', the cell discharges its granules into the restricted space between itself and the parasite. Because most of the MBP remains within this space it causes minimal damage to the body's own cells.

The role of mast cells and basophils in fighting parasites is more indirect: they do not appear to release substances that attack the parasite itself. Instead, they produce a variety of mediators, or chemical messengers, which stimulate a concerted attack by the immune system at the site of infection. Some of these mediators attract other cells, such as eosinophils. Other chemical mediators, including histamine and prostaglandins, make the blood vessels dilate and increase the permeability of the capillaries in the vicinity. These changes – which we observe as reddening and swelling, or 'inflammation'– allow other white blood cells to reach the site of infection more rapidly.

Mast cells can degranulate in response to various signals. Some of the complement products trigger them directly, adding to the inflammation produced by the complement proteins themselves. Indirect triggering occurs by way of a special type of antibody, known as IgE (immunoglobulin E), which is described in Chapter 19.

Although useful against parasites, mast cells can also cause allergies. This happens if the normal control mechanisms break down and the body begins making IgE antibodies in response to harmless substances, such as pollen or cat fur. The mediators released from the mast cells producc the allergic symptoms. In the case of hay fever, for example, there is intense inflammation of membranes in the nose and eyes during the pollen season.

Common origins
Ancient and modern

Despite their great diversity, the many cells of the immune system have a common origin in the body. They all develop from a single type of cell, the **pluripotent stem cell**. During their development, however, stem cells can follow one of two pathways. The first pathway gives a cell called a common

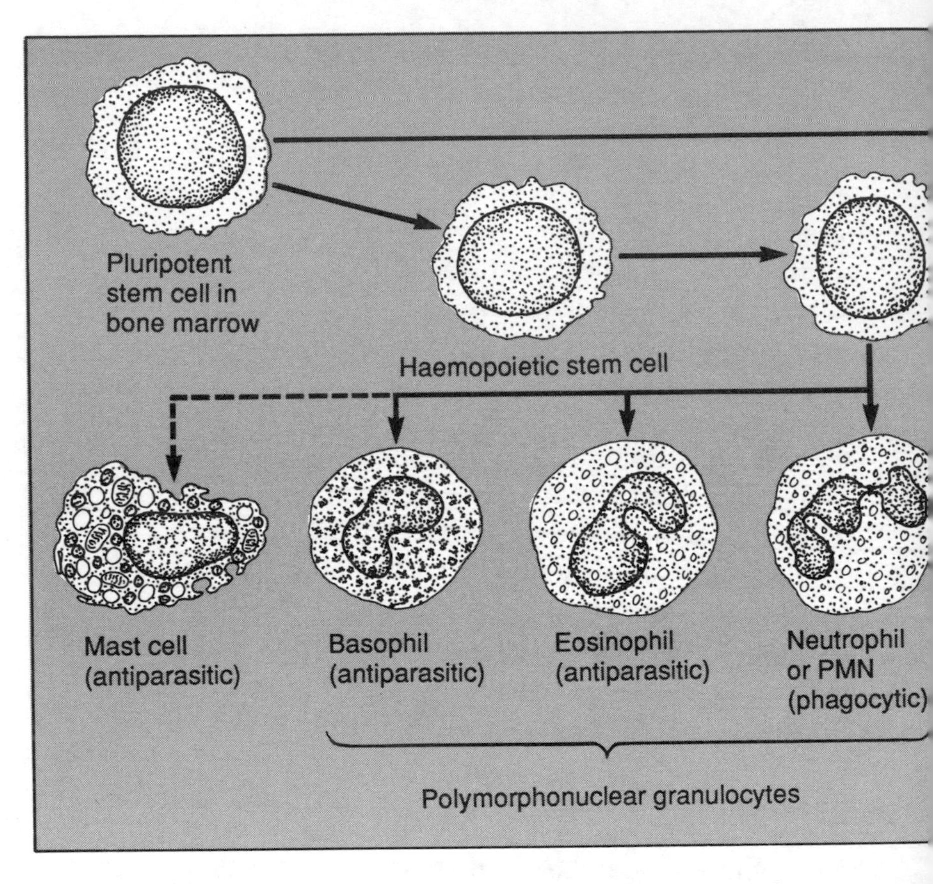

The many cells of the immune system all originate in the bone marrow from a common type of stem cell (pluripotent stem cell). As these stem cells divide, they can develop into myeloid stem cells or lymphoid stem cells

myeloid progenitor, which can then develop into a monocyte, PMN, eosinophil, basophil or mast cell – the more 'primitive' cells of the immune system. The second pathway gives a cell called a common lymphoid progenitor, which develops into a **lymphocyte**. Lymphocytes and the lymphatic system are the most recent evolutionary additions to our defences, but they work closely with the more ancient parts of the immune system.

The cells that develop from the lymphoid lineage are even

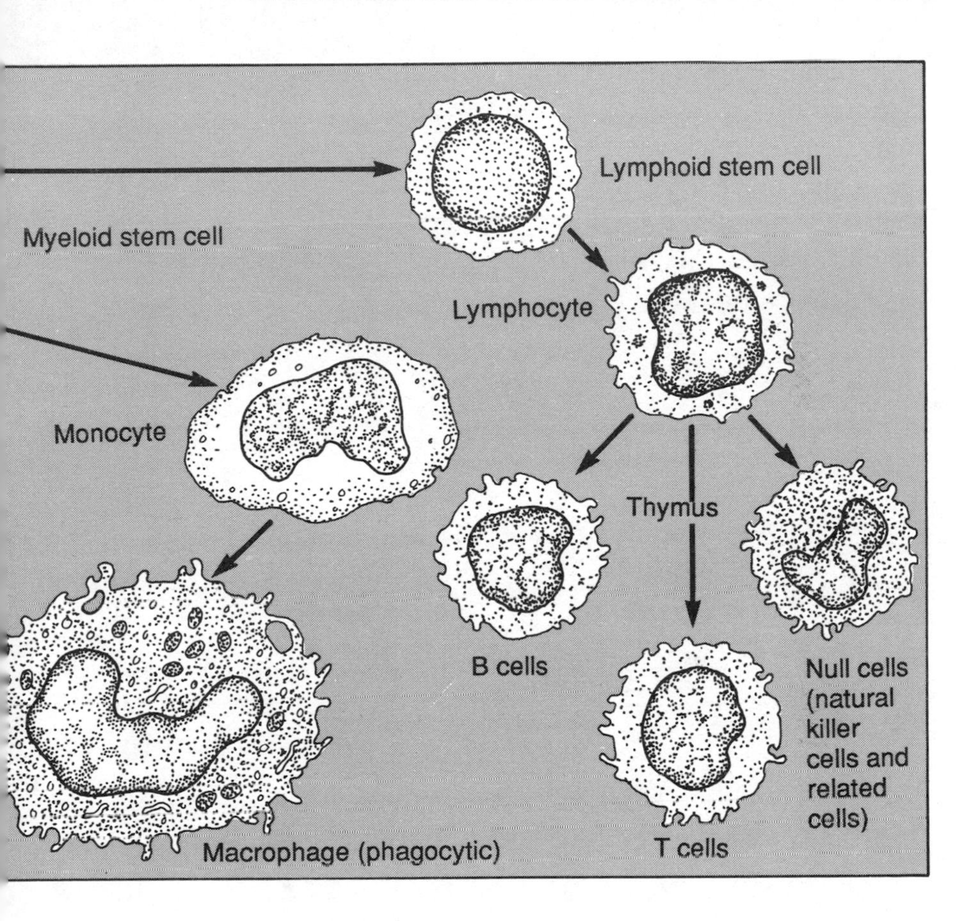

more diverse than those of the myeloid lineage in the ways they attack invaders. The three main groups are B lymphocytes (or **B cells**), T lymphocytes (or **T cells**) and null cells, of which the best known type are the natural killer cells (**NKs**). B cells are responsible for producing antibodies, and are very specific for their target (their 'antigens'). By contrast, NKs kill invaders directly, but are not specific. T cells cannot be summed up as easily as the other two groups, because they include at least four different types of cell with contrasting roles. However, they are all specific for their targets, just as B cells are.

This specificity, shown by B cells and T cells, is what marks lymphocytes out from the rest of the immune cells. The phagocytes and other 'primitive' cells can respond to widespread chemical markers, such as cell-wall components common to certain bacteria. But this system has distinct limitations. It fails to act against some types of bacteria, and it is not much use against viruses, which are very diverse chemically. The triumph of the lymphocytes, as we shall see in the next chapter, is that they can produce a highly specific receptor for *any* chemical marker.

10 March 1988

CHAPTER 19

The Human Immune System: The Lymphocyte Story

Linda Gamlin

The bite of the king cobra kills its victims within two hours by paralysing the muscles of the heart and respiratory system. Cobra venom is a cocktail of toxic substances, but the main ingredient is a neurotoxin – a peptide molecule that blocks messages from the nerves to the muscles. The victim dies unless injected with antivenin, which contains **antibodies** for the neurotoxin. Antibodies are specialized protein molecules that bind specifically to their target molecule, or **antigen** – in this case, the neurotoxin peptide. The usual means of producing antivenin is to 'milk' venom from a king cobra and inject it into a horse. The injections begin with a very small, non-lethal dose and gradually increase in strength. This stimulates the horse to produce antibodies to the neurotoxin. Blood taken from the horse will then yield a serum rich in antibodies. When injected into a human victim, these antibodies course through the bloodstream, combining with any neurotoxin peptides they encounter and thus inactivating them.

Snake bites illustrate both the strengths and the weaknesses of the antibody system. On the positive side, the body can manufacture antibodies to fit any chemical – neither the horse nor its ancestors need to have encountered a cobra before. This chemical feat is not achieved by producing 'tailor-made' antibodies to suit each new chemical as might be expected. Instead, the body generates a vast number of 'off-the-peg' antibodies, with an almost infinite variety of chemical 'shapes and sizes'. One of these is bound to fit any new antigen.

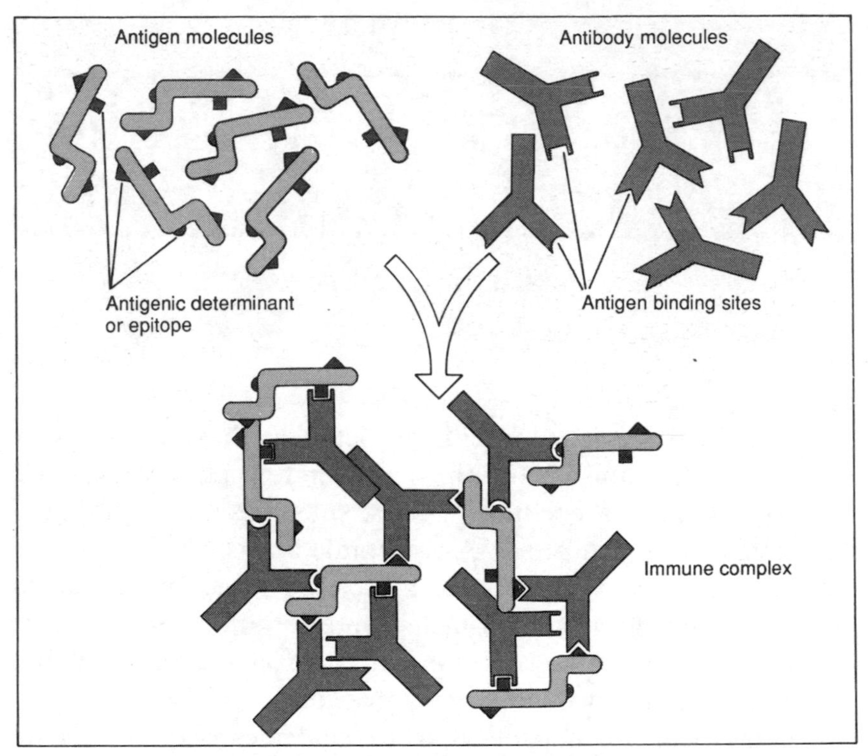

Antibodies combine with their antigen to form immune complexes. An antigen usually provokes several different antibodies – they each recognize different chemical features of the antigen. These distinctive features are called epitopes, or antigenic determinants

Antibody is produced by B cells, members of a group of cells known as **lymphocytes**, found only in the vertebrates. What makes lymphocytes special is their ability to produce highly specific protein receptors to combine with any chemical marker. The secret of their success is that they have multiple copies of the genes for the receptor proteins, each copy being slightly different. As the lymphocyte matures, these genes are recombined randomly to produce a new and unique gene for the protein receptor of that cell. In the case of B cells, the receptors are carried on the cell surface at first and are known as **immunoglobulins**. Only when they are released into the body fluids are they called antibodies. Every B cell has the capacity to produce large quantities of its own unique antibody, but it

The anatomy of antibodies

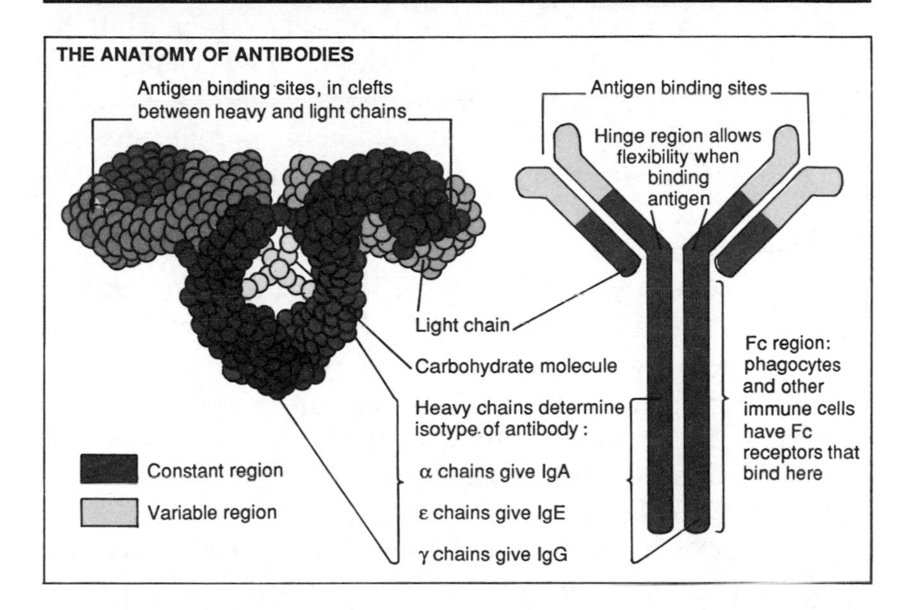

Each antibody molecule has four peptide chains: two heavy chains and two light chains. (On the left is a molecular model generated by computer, on the right a schematic diagram of antibody structure.)

All four peptide chains have a variable region, whose amino acid sequence is produced by genetic rearrangements. These variable regions create the antigen-binding sites. The Fc region can bind to phagocytes, and in some isotypes it can activate complement or make mast cells degranulate as well. Such isotypes produce more inflammation. B cells can switch from producing one isotype to another, depending on what sort of immune response is required.

---

does not do so until the antibody is needed. Pursuing the analogy of the clothes shop, with its vast selection of off-the-peg suits, B cells are rather like pushy sales assistants, urging any antigens they meet to try their particular immunoglobulin for size. If the B cell's immunoglobulin binds to the antigen, this has a dramatic effect on the B cell. Given the right signals from certain regulatory cells, the B cell divides rapidly and produces **plasma cells**. These are rich in protein-making apparatus and are, in effect, factories for synthesizing antibodies: one plasma cell can make 2000 antibody molecules a second. The process whereby an antigen promotes the production of a suitable antibody is known as clonal selection (the plasma cells it generates are genetically identical – clones).

Antibody diversity and clonal selection are a winning combination because they enable the immune system to respond to new challenges from previously unknown enemies. They are part of the **adaptive immune system**. Phagocytes and other 'primitive' cells (see Chapter 18) make up the **innate immune system** – they cannot mount an attack on novel organisms without assistance from lymphocytes.

So much for the strengths of the antibody system. Its weakness is that, with so many different B cells needed, there can only be a few of each genetic type. The body needs time to build up an effective stock of plasma cells from these few B cells, so it cannot produce much antibody initially. When challenged by a massive dose of snake venom the body may begin to produce antibodies, but it cannot do so fast enough to save the victim. In a disease such as measles there is a period of illness before the body musters enough antibodies to defeat the virus. Fortunately, during a bout of infection, memory B cells develop from the same clone as that which produces antibodies, and they survive in a dormant state for many years. If the same microorganism reappears the memory cells can proliferate and build up the level of antibody very quickly. They usually fight off the infection so rapidly that there are no noticeable symptoms. In other words, the person has become immune to that disease.

Vaccination creates memory cells artificially by various means. The principle is the same in all cases: to expose the body to toxins, or to antigens from the surface of the microorganism, but without any risk of damage or infection. This is done by heating the pathogen to make it non-infective, for example, or by inserting genes for its coat proteins into a harmless bacterium.

Versatile phagocytes
Fixing complement

The action of antibodies against snake venom is simple: the antibody alone disarms the toxin. Antibodies can also neutralize bacterial toxins, and combat some viruses by binding to them and preventing them from invading cells. But with the majority of microorganisms, the antibody needs help to kill the invader. That help comes from various sources.

Phagocytes carry receptors for the stem (Fc region) of antibody molecules. Binding to the Fc portion stimulates the phagocyte to engulf and digest the antibody and its associated antigen. Here the antibody is acting as a sophisticated type of receptor for the phagocyte, enabling it to recognize a far greater range of invaders. Some types of antibody are also able to 'fix complement', that is, to trigger the complement reaction. They do so by means of a complex set of reactions, known as the classical pathway, but the outcome is much the same as for the alternative pathway (see Chapter 18).

One type of antibody, IgE, has a special way of enlisting help against invaders, mainly multicellular parasites. This isotype binds by its Fc region to basophils and mast cells. If it then binds its antigen, the cross-linking of the IgEs stimulates the cell to release powerful mediators. These cause inflammation and bring other immune cells into the infected area.

Regulatory T cells
Helpers/suppressors

B cells are just one type of lymphocyte. The second major type, the T cells, are far more varied in their action and fall into four

The lymphatic system

Lymphocytes are so called because they are abundant in the lymphatic system, a network of vessels that drains fluid (lymph) from the tissues. The fluid derives from blood serum that oozes through the walls of the capillaries. Lymph contains no red cells and lacks many of the proteins of blood, so it appears pale and watery. But it contains many white cells because these, too, migrate through the capillary wall. Lymph eventually flows back into the circulatory system via the thoracic duct. As it travels through the lymph vessels it is filtered by the lymph nodes,

The lymph vessels and nodes (left) and the other lymphatic tissues (right)

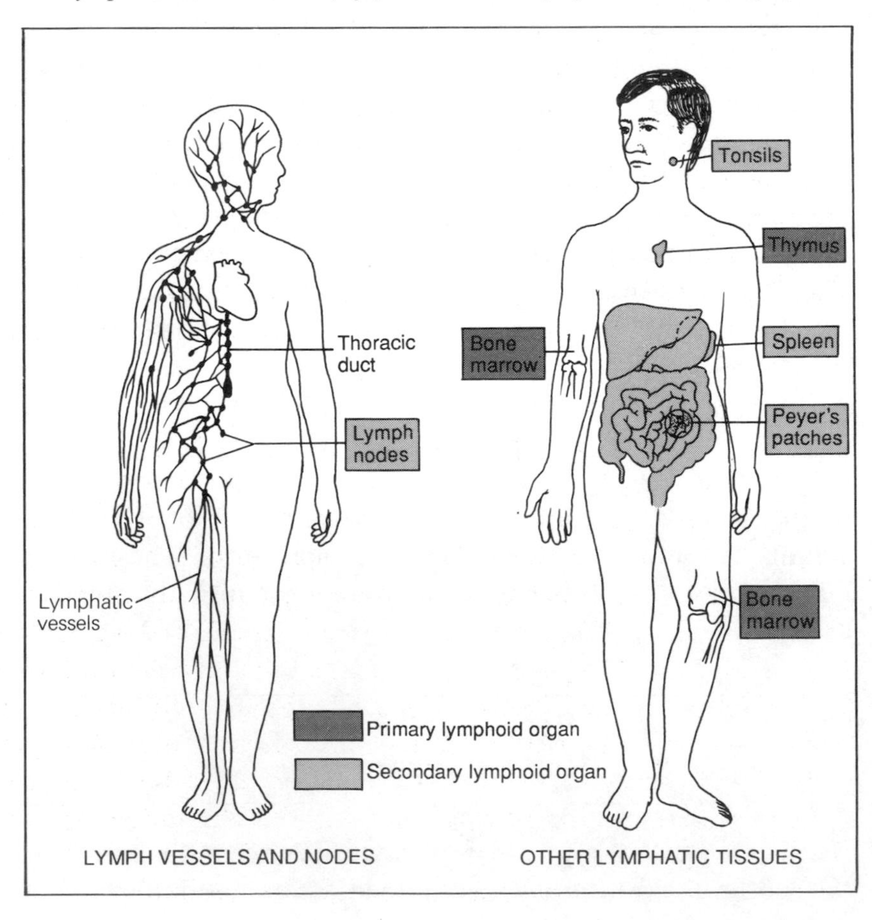

which contain large numbers of lymphocytes ready to fight infection.

Concentrations of lymphocytes also occur in other places, such as the tonsils, and areas in the gut wall known as Peyer's patches. These lymphatic tissues help to defend the areas of the body where infection can easily enter.

Lymphocytes all originate from stem cells in the bone marrow. Some then migrate to the thymus where they mature into T cells (the 'T' means thymus-derived). As the T cells mature an important process occurs, known as 'thymic education', which prevents the immune system from attacking the body's own cells. Exactly what happens during thymic education is still uncertain. Many immunologists believe that any Th cells with receptors that bind to the body's own molecules are destroyed. This process is called clonal deletion. Others disagree and believe the educative process may lie with the Ts cells, or elsewhere.

B cells mature in the bone marrow rather than the thymus, and autoimmune B cells are not eliminated. Because the T cells do not allow these autoimmune cells to proliferate, they normally produce no ill effects.

If lymphocytes are thought of as a police force, patrolling the body, then the primary lymphatic organs are the police training colleges, where they originate and learn their skills. The secondary lymphatic organs are the local police stations, where they congregate and deal with suspect antigens.

---

main groups. Two of these regulate the activities of B cells. **T helper cells**, or **Th cells**, stimulate B cells to divide and produce plasma cells. Their presence is crucial – a B cell cannot go into action unless it has 'permission' from a Th cell specific for the same antigen. **T suppressor cells**, or **Ts cells**, are the brakes of the immune system – they suppress the production of antibodies, probably by inhibiting Th cells specific for the same antigen. By regulating the activities of Th and Ts cells, the body can control what sort of antibodies it produces. Immunologists are not entirely sure how Th cells exert their influence over B cells. They may make direct contact and stimulate them to proliferate by means of localized chemical messengers; or, they stimulate them by means of other messenger molecules, called helper factors, with no direct contact required. Even less is known about how Ts cells work, but they may release suppressor factors.

Like B cells, T cells are highly specific for their antigen, thanks to receptor molecules on the cell surface. These are proteins that are chemically similar to immunoglobulins, but they have only two chains instead of four. The receptors are almost infinitely variable, as with B cells, and again, the T cells do not proliferate until they are needed – until an antigen binds to their surface receptor. They, too, are rather like sales assistants in a huge clothing store, urging antigens to try their particular receptor for size. But with T cells, particularly the T helper cells, the antigens are reluctant customers – they refuse to try on the receptor. Other 'sales assistants', known as **antigen presenting cells**, or **APCs**, must capture the antigens and force them to interact with the Th cells.

No analogy is perfect, and likening antigens to reluctant customers is slightly misleading. Antigens readily combine with antibodies, after all, so why should they not combine with receptors on a Th cell? The answer lies with the Th cell itself, which will only recognize the antigen in certain circumstances. An APC presents its antigen alongside a chemical marker known as a class 2 protein, and the Th cell must see the two together to respond. The class 2 proteins are just one of a set of protein markers, unique to each individual. Coded for by genes of the 'major histocompatibility complex', they are known as **MHC proteins**.

MHC proteins are a sort of chemical uniform that tells other body cells that they are friend rather than foe. In laboratory experiments Th cells from one individual will not usually cooperate with APCs from another individual, unless they are identical twins. This phenomenon, known as **MHC restriction**, affects other immune reactions as well. Limiting the reactions of Th cells, so that they only recognize their antigens alongside class 2 proteins, probably helps to keep them under control. If

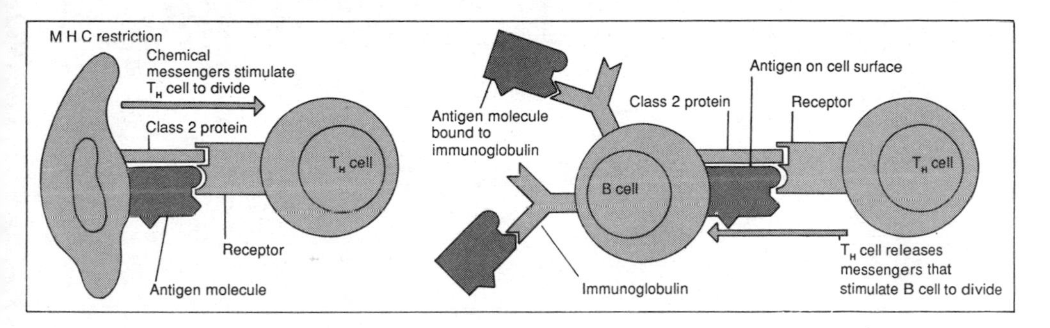

MHC restriction means that T helper cells need to see their antigen alongside class 2 proteins, which occur on APCs and B cells

they were unable to interact with the antigens except through an APC, this would localize the immune reaction in suitable parts of the body. It might also expose the Th cells to messages left by other lymphocytes on the APC – suppressor factors from Ts cells, for example. APCs could also carry messages of their own. Some immunologists suspect them of having an important role in discriminating between antigens and directing the immune response.

Tc cells and killers
Professional assassins

A third group of T lymphocytes, the **T cytotoxic cells**, or **Tc cells**, are professional assassins that kill cells outright. Their main targets are body cells infected with viruses, which they destroy before the virus can proliferate. Tc cells are also part of the defences against tumour (cancer) cells.

Tc cells have surface receptors that make them specific for a particular antigen. They bind to their target cell via this antigen, and then kill the cell with an enzyme that bores through its membrane. Like Th cells, Tc cells are MHC restricted – they must see their antigen alongside a characteristic protein marker. But it is not the same protein. Tc cells look for a class 1 protein, which is found on all body cells, not just on the APCs and B cells, as class 2 proteins are. With Tc cells, the reason for

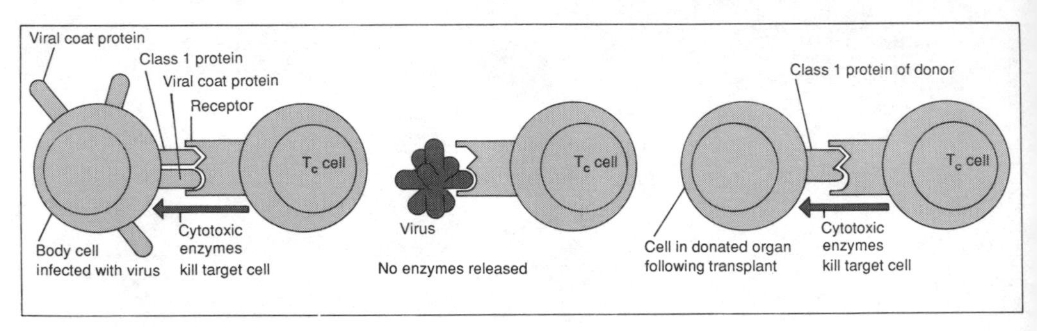

T cytotoxic cells need to see their antigen alongside class 1 proteins, which occur on all body cells

MHC restriction is clear. The Tc cell recognizes a body cell infected with virus by the viral coat proteins that begin to appear on its surface soon after infection. Because they need to see a class 1 protein alongside the viral protein, they are not distracted by futile skirmishes with free viruses, which they cannot kill. In the case of tumours, the membrane proteins of a cancerous cell probably change slightly, allowing the Tc cell to recognize it as abnormal. Unfortunately, Tc cells also attack cells from another individual, causing problems of rejection in transplant surgery.

Like B cells, Tc cells leave behind memory cells that can respond rapidly to a second infection. They are also regulated in the same way as B cells: Th cells stimulate them and Ts cells inhibit them. The importance of T helper cells, in both antibody production and Tc cell action, explains why AIDS is so deadly, and why it results in cancers as well as infections – the HIV virus invades the body's T helper cells and eventually paralyses them.

The third group of lymphocytes, the null cells, are also professional assassins. But although they resemble Tc cells, in killing cells directly, they are not specific for any particular antigen. There appear to be at least three different types, known as natural killer cells (**NKs**), lymphokine-activated killers (**LAKs**), and killer cells (**Ks**). It is not always clear how they recognize their target, although K cells have receptors for the Fc region of antibodies and can kill antibody-coated cells.

Lymphokines
Chemical war cries

The body has yet another line of defence against tumour cells and body cells infected with parasites. Macrophages can engulf and digest these cells – but only if they receive the appropriate instructions. The signal comes from lymphocytes, particularly T cells, that have recognized their particular antigen on the cell surface. In this excited state, the T cells release chemical mediators, known as lymphokines, which stimulate macrophages into a more aggressive form. Known as activated or 'angry' macrophages, they can engulf other cells much more readily. The angry macrophage is not specific for the antigen, but because the effect is local it tends to attack the cell recognized by the T cell's receptor. As well as fighting tumours, it is useful against fungal infections, and bacteria that have developed means of evading other immune cells.

Lymphokines have various other effects besides making macrophages angry (see Chapter 24). Some attract phagocytes to a site of infection, others stimulate their division, or prevent them from leaving the site. One particular group of lymphokines, the **interferons**, specifically affect body cells attacked by virus. They inhibit the protein-making machinery of infected cells, so that the virus cannot proliferate, and provoke changes in uninfected cells which make them resist the virus.

As immunologists discover more about the immune system, the important role of lymphokines is becoming clear. Every immune cell produces a range of these chemical messengers, which have multiple effects on other immune cells, and on the body as a whole.

24 March 1988

Further Reading

Immunology by Ivan Roitt, Jonathan Brostoff and David Male (Churchill Livingstone, 1989) is a textbook aimed at medical students but is accessible to a more general audience. *The Body Victorious* by Lennart Nilsson with Jan Lindberg (Faber, 1987) is a collection of superb photographs of immune cells.

CHAPTER 20

The New Genetics

Omar Sattaur

Imagine trying to find a needle in a haystack the size of Mount Everest. Impossible? Probably, yet scientists have made giant strides in molecular biology over the past decade enabling them to perform the molecular equivalent of that 'impossible' task.

Each cell in our bodies contains all the genetic information required to make an entire human being. That information is carried in the **genes** that make up our **chromosomes**. The stuff of genes is deoxyribonucleic acid (**DNA**), a spiral molecule resembling a ladder whose 'rungs' are built of pairs of **bases**. If we were to unravel the DNA of all the chromosomes in each cell in a person's body and join it end to end it would stretch to the Moon and back about 8000 times. Yet, with the new techniques of molecular biology, scientists can now isolate a single gene of perhaps 1000 to 2000 bases from an amount of DNA sufficient to contain more than six million genes of similar size!

Being able to locate and analyse a single gene promises to revolutionize the study of **inherited**, or **genetic**, **diseases**. It already enables doctors to diagnose some inherited diseases in human fetuses only eight weeks old. Examples are **haemophilia** and **sickle-cell anaemia**, two inherited diseases of the blood. The new techniques are also probing the mystery of how disorders such as cancer and heart disease develop. Eventually, they may even lead to **gene therapy**, in which people born with faulty genes may be given normal ones.

Inherited diseases
Genes from our parents

Genetic diseases follow two patterns of inheritance. Virtually all the cells in your body contain 46 chromosomes in their nuclei. The only exceptions are the **gametes**, the sperm and egg cells, which contain half that number. You will have inherited 23 of those chromosomes from your mother (via an egg cell) and 23 from your father (via a sperm). So all the genes on chromosomes come in two versions, one inherited from each parent.

In some cases a single abnormal gene from one parent is sufficient to cause disease; here the gene is said to be **dominant**. More commonly, a child must inherit a defective gene from *both* parents before the disorder shows itself; in this case the genes are called **recessive**. Healthy persons who have just one defective recessive gene are called **carriers**. They do not suffer from the disease but could pass it on to their children. We all carry a few potentially harmful recessive genes.

Some diseases, such as haemophilia, are caused by defective genes on the X-chromosome, one of the 'sex chromosomes'. Cells of females have two X-chromosomes whereas those of males have one X- and one Y-chromosome. A woman who has the haemophilia gene on both her X-chromosomes will suffer the disease. If she has the defective gene on one of her X-chromosomes, she will not show symptoms of the disease but is said to be a carrier. A male inheriting the defective gene on his X-chromosome will suffer the disease. Such disorders are said to be **sex-linked**.

Until very recently, doctors could not usually determine whether a fetus was afflicted with a serious genetic disease. All they could do was give some parents a rough idea of the risk of having an affected child. Such **genetic counselling** is usually feasible only if the parents are known to be carriers. This might be known because they already have one child with an inherited disease, such as **cystic fibrosis**. Children with cystic fibrosis make excessively thick secretions which can block the air passages of the lungs.

A doctor can tell the parents their chances of producing a second child with cystic fibrosis. This approach is limited by the fact that, in most cases, people do not know that they are carriers until they have produced an affected child. Nor are there any simple diagnostic tests to identify carriers of most genetic diseases.

For some diseases, doctors can now isolate and analyse the specific fetal genes suspected to be defective from as little as 2 millilitres of fetal blood, or two millionths of a gram of fetal DNA. The advances in molecular biology go hand in hand with new methods of obtaining fetal cells. Medical scientists developed the first method of obtaining fetal cells about 20 years ago. The technique, called **amniocentesis**, is carried out 14 to 15 weeks into pregnancy. Amniocentesis is used in **prenatal diagnosis**, in which doctors diagnose disease in a fetus. It gives the woman the chance to decide whether she wishes to continue her pregnancy.

In amniocentesis a doctor uses ultrasound to image the membranous bag that encloses the fetus within the womb. By means of a hypodermic needle the doctor draws up a sample of **amniotic fluid**. Amniotic fluid surrounds and cushions the fetus while it is in the uterus. The fluid contains some fetal cells which scientists can examine directly for chromosomal abnormalities. It also contains many chemicals involved in the fetus's metabolism. Scientists can detect a few rare inherited disorders of metabolism by measuring the quantity and turnover of these chemicals in amniotic fluid. They can now extract the DNA from the fetal cells and test for the presence of specific faulty genes.

The newest method of obtaining fetal cells, however, is called **chorion villus sampling (CVS)**. It allows doctors to make prenatal diagnoses of certain diseases after only 8 to 10 weeks of pregnancy. Amniocentesis can provide only a few fetal cells which doctors must then grow in culture to provide enough fetal DNA to analyse, and this can delay results by several weeks. In contrast, CVS can provide enough fetal DNA for direct analysis, and give an answer in a day or two.

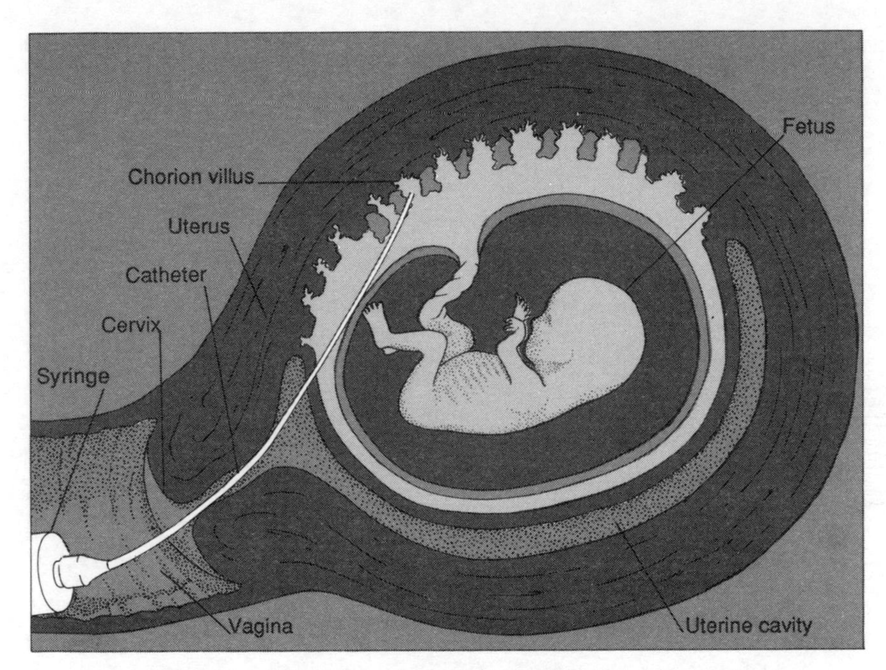

In chorion villus sampling, doctors can obtain fetal DNA from trophoblast tissue after only eight weeks of pregnancy

After the embryo has implanted itself in the uterus, the **chorionic plate** begins to grow around the fetus and later forms the **placenta** which feeds the growing fetus. The chorionic plate is made up of cells derived from the outer layer of the embryo. In CVS, doctors use ultrasound scanning to obtain chorion cells between 8 and 10 weeks of pregnancy. One sample can provide as much as 100 micrograms of fetal DNA.

New research in mammalian development suggests that scientists may, in the future, be able to carry out diagnostic tests in the 'pre-embryonic' period. This period spans the first two weeks after an egg has been fertilized and before it has begun to implant itself into the lining of the uterus. Using ***in vitro*** **fertilization (IVF)**, the pre-embryos could be checked for defects before replacing them in the woman's uterus. Such **preimplantation diagnosis** would offer women the chance of starting pregnancy knowing that their children will not inherit the disease in question, rather than finding out during the first or second three months of pregnancy. This is particularly advantageous for women who find abortion unacceptable for ethical or religious reasons.

Given sufficient fetal DNA, how can scientists locate the often small changes that give rise to inherited disease? This is where the new techniques of molecular biology come in. The

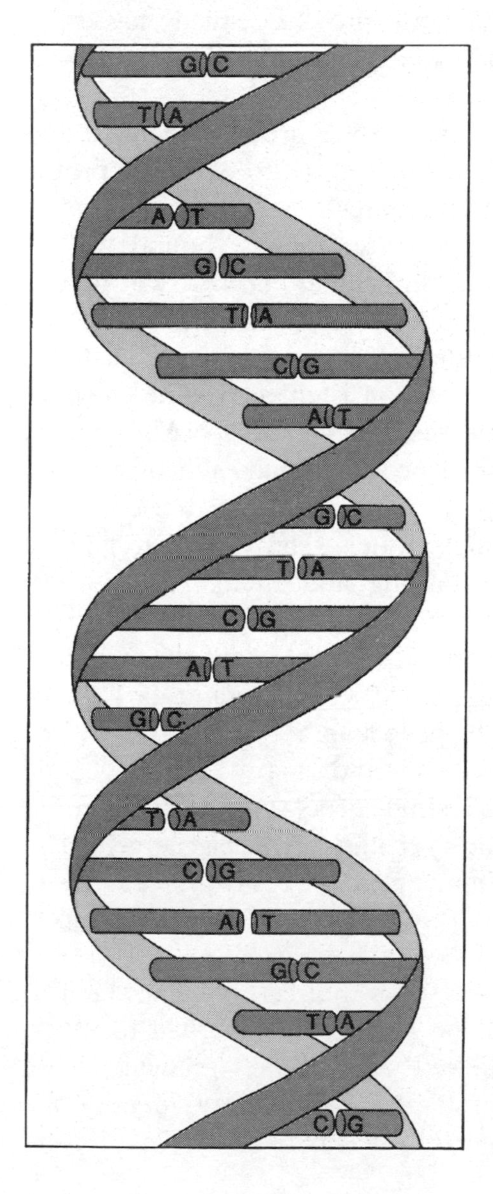

In its inactive form, DNA exists as a double helix bound together by specific pairing between complementary bases. During protein synthesis the two strands separate to reveal the base sequence from which messenger RNA is made

two component strands of DNA are bound together by specific pairing between so-called complementary bases – **adenine** binds only to **thymine**, and **guanine** only to **cytosine**. When one strand meets its match, they bind together. Scientists have exploited this fact of life to make **gene probes** for a number of mutations that give rise to genetic diseases. They isolated one particular intermediary between DNA and protein, called **messenger RNA (mRNA)** and then made a **complementary DNA (cDNA)** from it.

Exposed to cellular DNA the cDNA probe recognizes and binds to its complementary sequence. In preparing a probe, scientists incorporate radioactive bases that 'label' the cDNA and enable them to find it once it has bound to its complementary sequence in the cellular DNA. To find a particular gene in someone's DNA, scientists first chop the long strands of DNA into smaller fragments using bacterial enzymes that cut the DNA in a consistent way. The enzymes are called **restriction endonucleases**. Placing the DNA fragments on a special gel exposed to an electric field makes them separate according to size, and form a pattern. The pattern is then 'blotted' on to a nitrocellulose filter. The DNA fragments are bound to the filter by baking and then exposed to a radioactively labelled probe. When the filter is placed on an X-ray film the position of the probe, and therefore the fragment containing the altered base sequence, shows up as a dark band. The picture of bands is called a **gene map**.

Most haemoglobin disorders, and many other genetic diseases caused by a defect in a single gene, cannot be diagnosed in this way because we do not yet know which gene gives rise to the disease and so cannot make gene probes for the mutation. In such cases researchers may draw on another approach based on **genetic linkage**.

Researchers have known for years that certain genes tend to stay together on a chromosome and so are inherited together. The genes are said to be 'linked' as they are physically close together on a chromosome. If scientists cannot identify the product of a particular gene they wish to study, and if that

How our cells make proteins

The DNA in chromosomes is tightly coiled and folded. Stretched out, it resembles a long, spiral ladder. Each side of the ladder is built of a chain of sugar-phosphate units. In DNA the sugar molecule is deoxyribose. The rungs consist of pairs of the four bases: adenine (A), thymine (T), guanine (G) and cytosine (C).

Because of the chemical properties of the bases, A pairs only with T, and C only with G. This means that the two complementary strands cannot fit together in any other way.

Proteins are built of chains – called **peptides** – of **amino acid** building blocks, each chemically linked to the next. Certain chemical properties of the amino acids force the chain to fold in particular places, producing a three-dimensional structure whose shape is critical if the protein is to perform its job properly.

The information that directs the order of amino acids in peptide chains is contained in the sequence of bases in DNA – the **genetic code**. A gene is a length of DNA that codes for one peptide.

The genetic code is read in triplets of bases in the DNA molecule. In other words, each amino acid is represented by a code word of three bases; for example, the amino acid valine has the code word GTG. A triplet of bases that represents one amino acid is called a **codon**.

When a cell needs to make a particular protein, something has to 'read' the relevant genes and translate each codon. Then the appropriate amino acid must be found, and all the amino acids must be assembled together in the correct order to make the protein.

The first part of this process goes on in the cell nucleus, the home of the DNA. An enzyme called **ribonucleic acid polymerase** copies one of the strands of DNA to form a molecule called **messenger RNA (mRNA)**. This process is called **transcription**. RNA is similar to DNA except that its sugar is **ribose** instead of deoxyribose and it contains the base **uracil (U)** instead of thymine.

The molecule of mRNA then travels from the nucleus to the cytoplasm of the cell. There it acts as a template upon which the amino acids are joined

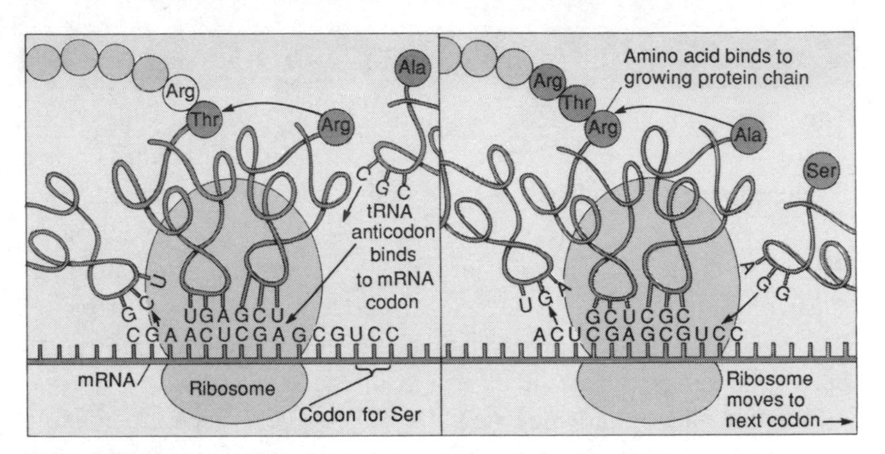

Messenger RNA is the template upon which proteins are built. Each codon represents a particular amino acid and is matched by an anticodon on the tRNA

together. This event is called **translation**, because the codons of mRNA are translated into the amino acids they encode.

Amino acids are brought to the appropriate place on the mRNA strand by molecules called **transfer RNAs (tRNAs)**. Each tRNA has a triplet of bases, called an **anticodon**, which finds and joins up with its complementary codon on the mRNA strand. In this way amino acids are placed in the appropriate order as directed by the sequence of mRNA codons.

---

gene cannot be located directly by a gene probe, the next best thing is to find a gene that is linked to it but *can* be located. Scientists can use such neighbouring genes as 'markers' to look for the presence or absence of the gene they wish to study. One problem is that few marker genes for genetic diseases have so far been found. In addition, the marker may not be close enough to the gene to be accurate. Recombination – where paired chromosomes physically swap parts of themselves – can break up the linkage between a marker and a disease-causing gene. Family studies are required to trace the inheritance of the marker and faulty gene through several generations.

Nonsense DNA
New marker genes

Researchers have since discovered non-coding regions of our DNA that contain base changes which either produce new cutting sites for restriction enzymes or remove existing sites. These have provided them with new markers. These harmless base changes follow the usual pattern of inheritance. The sequences are said to be **polymorphic**, that is, they occur in several different forms. Those that change the cutting sites of restriction enzymes are called **restriction site polymorphisms**.

Linkage analysis

Consider two parents who are both carriers (heterozygotes) for a deleterious gene **A**. Each parent has a chromosome that carries gene **A** and another that carries its normal counterpart **N**. If the mutation that produces **A** cannot be identified directly, linkage analysis may help.

In this example, one of the parental chromosomes carries a polymorphic restriction enzyme site **P**, which is close enough to the loci **A** and **N** that they will not be separated in successive generations. Here, the chromosome containing the polymorphism is designated + and that which does not −.

From the − chromosome a restriction enzyme cuts out a piece of DNA 10 kilobases (kb) long, containing locus **N** for which there is a gene probe. (1 kb = 1000 bases.) The + chromosome, however, contains a single base change that produces a new site for the restriction enzyme at **P**. Hence the fragments containing locus **M** from the + chromosome are now only 7 kb long.

Mapping the parents' DNA with probe **M** produces two bands, each representing either

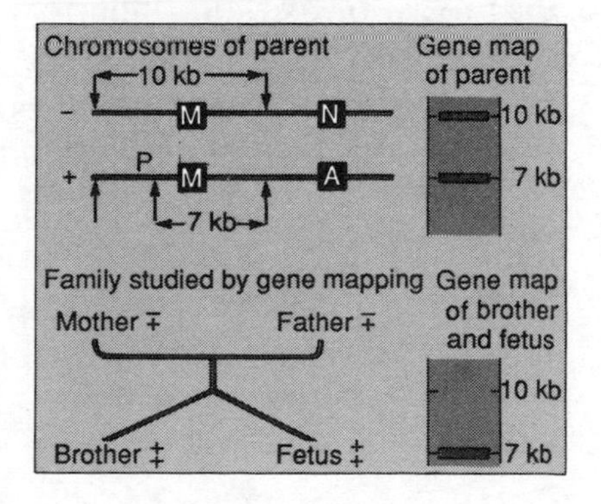

the + or − chromosomes. But which chromosome carries gene **A** and which one carries the normal gene **N**?

That answer comes from a family study. A previously born child has received gene **A** from both parents and its gene map reveals the + + chromosome arrangement; that is, only a single 7 kb band from mapping with probe **M**. For prenatal diagnosis in the next pregnancy, doctors can similarly map the fetal DNA with probe **M**. In this example, the gene map of the fetus also shows the + + arrangement and so it must also have received the defective gene **A** from both parents.

The length of the fragments can also be altered by the presence of varying numbers of small, repeated sequences of non-coding 'nonsense DNA', called **hypervariable regions (HVRs)**. HVRs occur throughout our DNA and are also inherited in the normal way. Because both HVRs and restriction site polymorphisms affect the sizes of DNA fragments they are collectively called **restriction fragment length polymorphisms (RFLPs)**.

Diseases of the blood
Rogue haemoglobins

The new techniques have spectacularly improved prenatal diagnosis of certain inherited diseases of **haemoglobin**, the protein that makes blood red and enables it to carry oxygen from the lungs to the tissues. Some people have imperfect haemoglobin genes. These 'faulty' genes can cause a range of diseases in which haemoglobin fails to do its job properly. In one of those diseases, called sickle-cell anaemia, the haemoglobin forms long, rod-like filaments that change the shape of the person's red blood cells when they give up their oxygen to tissues. This, in turn, causes the cells to be prematurely destroyed and to clog small blood vessels. The blockages cause severe illness which can be life-threatening.

Knowing the sequence of the haemoglobin gene, and studying the patterns of inheritance of sickle-cell anaemia in affected families, enabled researchers to work out the genetic mutation that gives rise to the disease. Haemoglobin is made up of two pairs of peptide chains called **alpha** and **beta** chains. The structure of those chains is controlled by corresponding **alpha** and **beta** genes. Sickle-cell anaemia is due to a single base change in the sixth **codon** of the beta chain gene, where each codon specifies a particular amino acid in the protein product. Instead of GTG, the codon for valine, the triplet reads GAG which codes for glutamic acid. That mutation of adenine for thymine – and the consequent substitution of glutamic acid for valine – changes the three-dimensional structure of haemoglobin. Once scientists knew the mutation it was possible to make a gene probe that would identify the mutation in a mixture containing the entire DNA of thousands of cells.

The discovery of RFLPs gives researchers the opportunity to study and, hopefully, diagnose genetic diseases in which the underlying cause is still unknown, such as cystic fibrosis. Suppose doctors want to find out if a fetus has received a gene for cystic fibrosis from each of its parents. They know roughly where on the chromosome the mutation is but are unable to identify it. The idea is to find an RFLP close enough to the faulty gene to act as a linkage marker for it.

Doctors examine the DNA of both parents and a previously affected child or other relative for a linkage between an RFLP and the gene carrying the cystic fibrosis mutation. If they find a linkage, they then look for the presence of the linked RFLP in the DNA of the fetus. Should the fetus show the presence of the polymorphism on *both* chromosomes, it must have received the mutant gene from both parents and thus would be expected to develop the disease.

Tomorrow's genetics
A window on diabetes?

So far these new techniques have been used to study diseases caused by single-gene mutations. What use could they be for studying other diseases, such as coronary heart disease?

Doctors know that many genes play a part in the degeneration of the coronary arteries that supply the heart muscle with blood and can lead to heart attacks even in young people. The involvement of many genes, plus their interaction with the environment, presents a much more complex problem to solve: heart attacks do not run in families in the same way as haemophilia or sickle-cell anaemia. Even so, the new techniques may help to clarify what goes wrong in such diseases and may eventually enable us to diagnose those people with a tendency

Gene therapy: hopes and fears

Gene therapy, to replace a missing or defective gene, may in the future cure many inherited diseases. The most likely candidates for gene therapy are single-gene disorders such as the haemoglobin diseases and disorders of metabolism arising from single-base mutations, because they are the most well-studied genetic diseases. For example, it may prove feasible to correct defects in the precursors of red blood cells that are made in the bone marrow. Although plans are afoot to begin gene therapy for people who would otherwise die from genetic disease, there are still many difficult problems to tackle.

First, there are still many obstacles to placing a gene in the chromosomes of each one of millions of host cells. Secondly, getting the gene into a cell, or even integrated into the host cell's DNA, does not mean that the gene will do its job. The regions of DNA that flank a gene are known to contain base sequences that are crucial in deciding whether the gene will be 'read' and in regulating the rate of its transcription. Finally, the additional gene must not interfere with any other processes in the normal life of host cells.

At present, animal experiments are providing some insight into solving these problems but it will be years before they are satisfactorily answered. However, researchers are putting a lot of effort into persuading viruses to solve the first problem for them.

Certain viruses, called retroviruses, insert their own genes into the chromosomes of the human cells they naturally infect. Although retroviruses cause serious diseases, including cancer, it is possible to disable them with the new techniques of molecular biology. Their disease-causing genes can be removed and a therapeutic gene 'stitched' on to the gene that enables the virus to enter cells. Some experiments in animals show that the technique is feasible.

The idea of introducing foreign genetic material into a human being can be frightening. If gene therapy became possible would it mean that we could choose the eye colour, musical ability or sex of our children? Would that be desirable or ethically acceptable? Even if gene therapy were used as a last resort to treat only life-threatening diseases, could doctors guarantee that it would be safe and not lead to unforeseen diseases such as cancer?

In June 1988 *The Lancet*, one of the world's most important medical journals, published a joint statement concerning gene therapy in human beings, by the medical research councils of a number of European countries, including Britain. The statement made it clear that gene therapy should be used only to correct genetic defects and that 'attempts to enhance general human characteristics should not be contemplated'.

It said that gene therapy should be carried out on body cells (**somatic cells**) only, and never on germ cells (sperm and egg). This means that any gene therapy would benefit only the individual treated and not his or her children. 'Further technical improvements in the expression of transferred genes in somatic cells will be necessary before successful gene therapy can be achieved even in animal models; in the meantime trials in man are not justified,' the report said. It recommended that countries form groups to consider proposals for gene therapy and to oversee trials.

to develop heart disease. One approach would be to isolate genes that might be involved in the chemistry of the arterial wall and in the way in which the body uses fats. For example, several genes alter the behaviour of the lipoproteins, which carry the fat that is deposited in the walls of blood vessels. As the vessels become narrower the risks of heart attacks increase.

Linkage studies will help researchers to identify the most important genes predisposing people to heart disease. Scientists can analyse families to see if particular genetic markers are associated with premature heart attacks. The same approach could be used for other diseases, such as diabetes.

3 December 1988

Further Reading

The New Genetics and Clinical Practice by David Weatherall (Oxford University Press, 1990) is an account for the non-specialist of what the new techniques of molecular biology offer medicine now and may offer in the future. *Human Gene Therapy* by Eve Nichols (Harvard University Press, 1988) gives an accessible account of state-of-the-art gene therapy and how it may develop. *Unnatural Selection? Coming to Terms with the New Genetics* by Edward Yoxen (Heinemann, 1986) is a readable introduction to the biology of prenatal diagnosis and gene therapy and the social and political issues they raise.

CHAPTER 21

The Nervous System: Getting Wired Up

Georgina Ferry

The nervous system – the brain and spinal cord, and the nerves that connect these to the muscles and organs – is the control centre of our body, allowing us to interact with the outside world and to keep in touch with the needs of the world enclosed by our skin. It carries information from our senses, which the brain interprets in the light of past experience; the brain then responds by generating commands, which pass through the nerves to the appropriate parts of the body.

The whole system is immensely complex, involving thousands of millions of nerve cells, or **neurons**. Each of these neurons may be linked with up to 1000 other cells at connections, or

Signals pass from neuron to neuron by chemicals released at the synapse

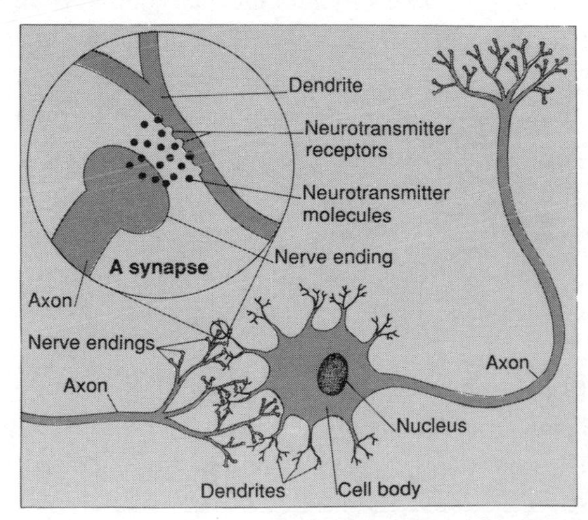

synapses, where they communicate with each other by releasing chemicals called **neurotransmitters**. For an animal to find food, a footballer to score a goal, or a scientist to explore the physics of the Universe, these neurons need to be organized in precisely connected networks. Even a touch on the hand, for example, involves a signal being passed along **sensory neurons** in the arm, into the spinal cord and up to the brain, with the signal passing, like a relay baton, from the synapse on one neuron to another at least three times. It arrives in a region of the brain that deals only with touch, which interprets the signal and communicates with neurons in the area of the brain concerned with movement. This will respond to the touch by sending out appropriate signals, perhaps to withdraw the hand, via the **motor neurons**.

The neuron – the basic unit of the nervous system

Nerve cells, or **neurons**, consist of a cell body containing the nucleus and other cellular machinery, usually with one long fibre, or **axon**, to carry their electrical impulses to other cells.

Attached to the cell body is a mass of threadlike extensions – **dendrites** – to receive incoming messages. At its tip the axon also branches into a number of terminals, or **nerve endings**. These form junctions, **synapses**, with the dendrites or cell bodies of other cells.

The nerve cells communicate by releasing chemicals called **neurotransmitters** at their synapses. These cross a tiny gap and bind to **receptors** on the membrane of the next cell. This has the effect of either increasing or decreasing the electrical excitability of that cell. If enough molecules of neurotransmitter bind to a cell, it fires an impulse, a wave of electrical activity that passes along the length of the axon.

When the impulse arrives at the nerve endings, they release their packets of neurotransmitter, thereby passing on information in their turn. The simplest of our thoughts or actions probably requires millions of these interactions to take place.

How are such networks built up? Does the set of instructions contained in the genetic material of each nerve cell direct the connection of every last synapse? Or does the position of each neuron and the synapses it forms depend on other influences at work in the developing embryo? Of course, genes are important. However, it would be impossible for the genetic code, powerful as it is, to carry *all* the information involved in the huge task of wiring the body. We now know that, at each stage of development, neurons receive signals from their environment which tell them where to go and what to do.

By the time we are born, our full complement of nerve cells is already in place. We will never acquire any more, and those that are lost, through disease, injury or age, will not be replaced. This does not mean that the system is complete at birth. There is still a lot to do to refine the connections between neurons. These connections will continue to be modified, possibly throughout our lives. This modifiability, or **plasticity**, is essential if we are to form new memories and learn new skills. But it is limited to the fine detail of the connections, and the flexibility gradually declines as we grow older. If the brain is damaged after birth there is little or no scope for rewiring to compensate for the damage. One aim of studying the initial development of the nervous system in the embryo is to discover what sets these limits to the growth of mature neurons.

From egg to embryo
The first nerve cells

When a sperm fertilizes an egg cell the egg divides first into two, then into four, and so on until a new individual forms. To begin with, the embryo is a simple ball of apparently identical cells. At this stage, in mammals, each cell has the potential to give rise to any of the many tissues and organs that make up our body. A key question for those studying embryonic development is at which level of subdivision a cell becomes committed to a single option.

The first sign of certain cells being earmarked for a future in the nervous system occurs at about 10–14 days in the human embryo. A dimple appears in the surface, which gradually deepens, as if someone were pressing a finger into a soft rubber ball. At this point, the infolded layer of cells, known as the **mesoderm**, lies under part of the outer, or **ectodermal**, layer. Chemical signals pass from the mesoderm to the ectodermal cells, instructing the latter to begin the construction of the brain and **spinal cord**. From now on, these ectodermal cells are irreversibly committed to being nervous system cells, as well as to the region along the body's axis they will occupy.

They start by forming a structure known as the **neural plate**, a sheet of about 125 000 cells. This rapidly undergoes dramatic changes: it elongates and forms a groove down the centre, whose sides roll up to form a tube. The **neural tube** closes first in the area that will become the neck, and then progressively upwards and downwards. It becomes detached from the overlying ectoderm, and develops into the spinal cord – the main thoroughfare for communication between the brain and the rest of the body. Meanwhile, three swellings appear at the head end, which are the primitive subdivisions of the brain. In human beings, all this has taken place by the end of the fourth week after conception.

Scientists still do not fully understand how the primitive shapes of body and organs develop from a simple ball of cells. Genes obviously decide whether the embryo will turn out to be a frog or a rat or a human being, although at this stage the embryos all look remarkably similar. But just as the formation of the neural plate depends on contact between the mesoderm and the ectoderm, it seems that local interactions encourage cells to congregate so as to form these rudimentary shapes.

A recent advance is the discovery of **cell adhesion molecules (CAMs)** distributed on the surfaces of cells. These control the stickiness of one cell to another, or of cells to the substrates – surfaces – on which they are growing. There are a number of different CAMs, each of which may stick exclusively to one substrate, so restricting the pathways open to the cells that

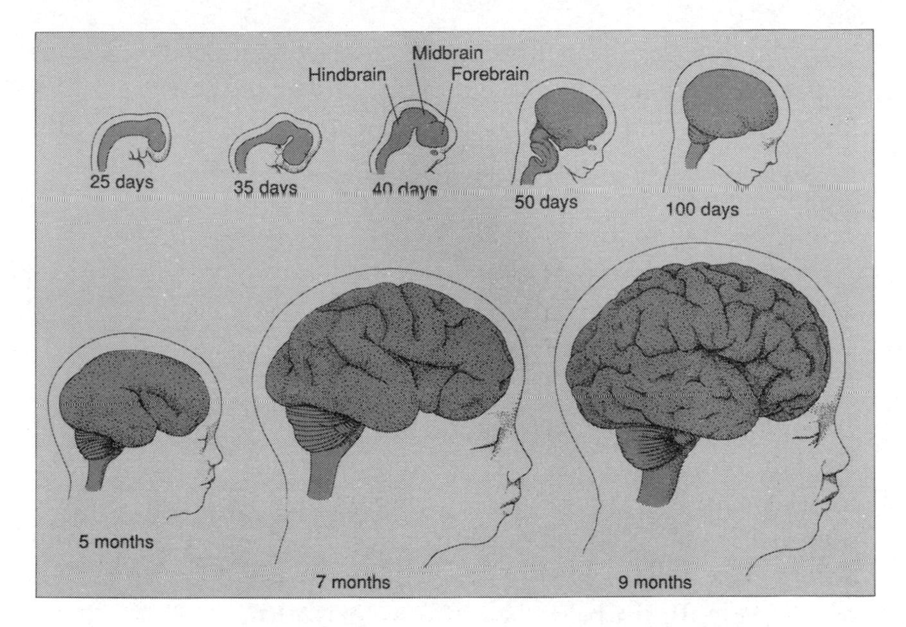

Development of the human brain (above). Until about 10–14 days, the embryo consists of apparently identical cells. During the process of gastrulation (so called because it results in the formation of the embryonic gut), cells in the ectoderm begin the task of constructing the nervous system. They start by forming the neural plate (below). This rolls up to form the neural tube from which develops the spinal cord

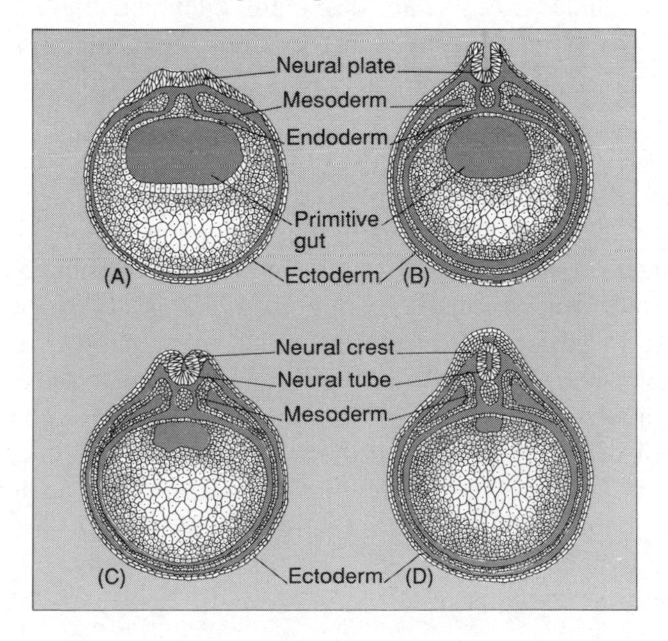

bear them. One, the neuronal cell adhesion molecule, is concentrated in the nervous system. The number of these molecules increases and diminishes at different times during the early development, increasing or decreasing a cell's chances of changing its position.

Differentiation
Who does what

Once the neural tube has closed along its entire length, its cells begin to reproduce themselves rapidly. They divide on the inner surface of the tube, and the daughter cells migrate outwards. Following migration, the cells stop dividing and **differentiate** – that is, they take on the specialized identities that will fit them for the roles they have to perform.

The mature nervous system contains many different types of nerve cell. Some are large, some are small; some are long and spindly, others round and densely branching: according to type, they use one or more of the few dozen neurotransmitters so far identified. A cell's size, shape and chemical characteristics all help to define its role. What, then, influences a newly formed cell to become one particular type of neuron? A variety of experiments, such as those of the French embryologist Nicole Le Douarin, have led to the conclusion that the **environment** in which an undifferentiated cell finds itself is a critical influence. For example, a cell may detect a chemical whose concentration increases from one end of the embryo to the other. Once in position, a cell may also change its properties in response to patterns of electrical activity or chemicals secreted by its neighbours. These and other interactions with the environment may determine the neuron's final identity through switching on or off the genes for producing transmitter chemicals, cell adhesion molecules or proteins for building new membrane. Exactly how they do this is largely unknown. The result, though, is usually permanent: the scope for change in a fully differentiated nervous system is very small.

Where you are, not where you came from

The French embryologist Nicole Le Douarin carried out a series of ingenious experiments showing that at the point when the basic plan of the nervous system is being laid out in the embryo, its cells still keep their options open. She took advantage of the fact that the nuclei of cells from a quail embryo, stained for observation under a microscope, looked very different from those of a chick. This meant that she could always trace quail cells transplanted into a chick even if they had moved away from the site of implantation.

Le Douarin transplanted a section of the neural tube from a quail embryo to the neural tube of a chick embryo. The cells transplanted included those of the **neural crest**, a region that originates in the neural plate but which is left outside the tube as it closes. Some of these cells go on to form the **autonomic nervous system**, which controls the internal environment by monitoring such things as heart rate and the composition of the blood, and taking whatever action is necessary to adjust them. It is also responsible for the changes we recognize as belonging to the 'fight or flight' response to danger.

In the autonomic nervous system there are two branches, called the **sympathetic** and the **parasympathetic** nervous systems, which have different but complementary functions. The nerve

The position to which cells are transplanted affects the type of neurons they develop into

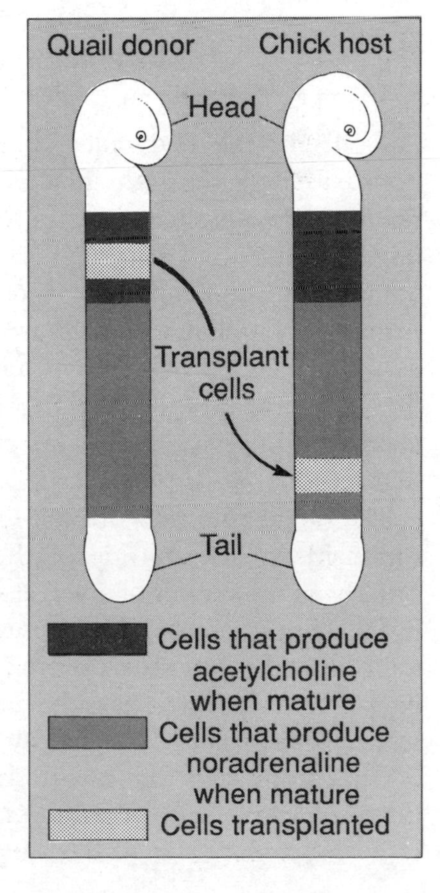

cells in each system use a different neurotransmitter to exert their effects: in the parasympathetic system it is **acetylcholine**, and in the sympathetic it is **noradrenaline**.

Le Douarin took cells from the neck region of a quail embryo and transplanted them to a position near the tail in a chick. She found that when the cells matured, those that would have produced acetylcholine if they had stayed in the neck, produced noradrenaline when transplanted.

In other words, in adopting their final identities the cells took their cue from the environment in which they found themselves.

Getting there
Movement of nerve cells

All cells in the body face the task of getting themselves to approximately the right place to carry out their functions. A neuron, though, must take up the right position with respect to its neighbours and also send out a long fibre, or **axon**, to make contact with appropriate target cells. Both tasks require feats of navigation that scientists are only beginning to understand. Newly generated cells move outwards from the inner surface of the neural tube to its outer layers. For some of these cells a scaffolding is already in place to guide them to their destinations.

The **radial glial cells** put out long thin fibres that stretch from the inside to the outside of the neural tube, in a pattern as orderly as the spokes of a wheel. Once the business of constructing the brain is complete, these glial cells disappear or change their function. Migrating nerve cells that are destined to become part of the cerebral cortex, the outer part of the brain responsible for conscious experience, simply follow the radial glial fibres.

Once in place, a cell has to make contact with others. Most types of nerve cell grow a mass of branching **dendrites** to receive incoming signals from other cells. Different types of cell grow dendritic 'trees' of characteristic shapes, but the precise arrangement of the dendrites depends on a cell's inputs. The neurons also grow one (or less frequently two) axons to transmit signals. In the embryo, the axons, which are only a few micrometres across, may have to navigate distances of several centimetres – for example, those that connect the spinal cord with muscles in the toes. The growing axons have specialized structures at their tips, called **growth cones**. These can extend and retract threadlike extensions as well as broader sheets of cytoplasm, and are packed with the cellular machinery for making new tissue. The axons extend by adding this material just behind the growth cone. So, how do these axons know where to go, and what tells them to stop when they arrive? We do not know the full explanation, but almost certainly they are in constant touch with their surroundings, looking for signposts to help them on their way. Growing axons seem, literally, to feel or sniff their way along.

An idea that has been very influential is that of **chemo-specificity**. The Nobel prizewinner Roger Sperry proposed that each cell had its own chemical identity, as well as the chemical 'address' of its ultimate destination. The idea is still being tested. Recent laboratory experiments suggest that the concept should be broadened to include a wide range of processes that may be at work to guide growing axons.

When immature nerve cells are grown in glass dishes in the laboratory, a number of factors influence their direction, which could also be at work in the living body. They tend to send their axons along scratches in the glass dish, suggesting that they look for preformed pathways or lines of least resistance. They prefer to grow on adhesive surfaces, and there are various proteins in the body that might provide differentially sticky routes through the tissues, acting in conjunction with the adhesion molecules on the cell's surface. They are attracted or repelled by electric fields of different polarity; such fields exist naturally in the developing embryo. Lastly, they turn towards nearby sources of body chemicals such as nerve growth factor (see below).

Overall, it seems likely that all these operate at different times and places; for example, differential stickiness may be important in getting an axon into the vicinity of its destination, while local chemicals may be more important in helping it to home in on the right target.

Cell death
Only the fittest survive

Between 30 and 75 per cent of all cells generated in the nervous system die by the time the system is complete. This seems rather wasteful, but it makes good sense in an organ of such complexity as the brain. Cell death in development is a way of matching the number of neurons to the needs of the developing brain. Removing one limb from a chick in the egg before the axons of motor neurons arrive leads to many more cells dying than normal. On the other hand, grafting on an extra limb bud means that a larger number survive. It seems that the cells compete for limited supplies of a **growth factor** provided by the target. Like wild birds competing for scarce food in winter, those who lose out will die. Scientists assume that there are many specific growth factors (a term that encompasses any chemical that a cell needs to grow or survive) but have identified only a handful so far. The best known is a protein called **nerve growth factor**, or **NGF**.

The sensory neurons that bring information from the skin and organs to the spinal cord cannot survive without NGF. In addition, they will not grow axons if kept in isolation without either pure NGF or some of their target tissues. Rita Levi-Montalcini, an Italian cell biologist, discovered in the 1950s that a cluster of such cells isolated in a glass dish and provided with NGF quickly produced a dense halo of fibres.

As well as regulating the number of surviving cells, competition ensures that axons that have grown in the wrong direction will be among those weeded out, because they will fail to reach a source of the particular growth factor they need. Some physiologists think that competition between cells is important in the early development of the nervous system. Such a concept is helpful, for example, in explaining how neurons become synaptically linked into circuits and networks as memories and skills are learned. In this example, though, rather than whole cells being eliminated, it is probably a matter of competition between individual synapses.

After birth
Time for fine tuning

Exactly how neurons recognize and make contact with their opposite numbers at synapses is almost entirely unknown for most parts of the nervous system, but some form of chemical recognition is almost certainly involved. The precision with which they do this can be remarkable, but it still leaves some scope for modification by experience. For example, in some young animals unable to see with the right eye, the cells that would normally respond to the left eye expand into the right's territory. Basic research shows that such changes can take place – or be reversed – only during a 'critical period' in early infancy. Assuming that the same applies in human infants, squints or other forms of poor vision need treatment early, before the pathways become irreversibly fixed.

Of course, human behaviour is remarkable for its flexibility – we can learn, either from our elders and betters, or directly from experience. This flexibility must have a direct counterpart in the brain, where memories and skills are stored in the activity of networks of nerve cells. Scientists now believe – although proving it is not easy – that the synaptic connections between cells in these networks are constantly modified in response to the activity of the network as a whole.

When it comes to compensating for serious damage, such as a head injury that affects the brain, the nervous system has a more difficult problem. If their target cells die, fully developed neurons can make new connections only up to distances of a few micrometres, while cells in the brain or spinal cord whose axons have been severed cannot grow new ones. Cells that lose their connections eventually degenerate. For someone who has lost skills, such as speech or the ability to make a cup of tea, the only hope lies in training the brain to reach the same goals by using other routes. Success in this kind of rehabilitation depends on the skill of the trainers – it is unlikely to happen of its own accord.

Why should the most important part of the body be so helpless in response to damage, while skin, bone and other tissues heal themselves with little trouble? It seems that all the special signals that support the growth of nerve cells and guide them during their development disappear as soon as the system is complete, and without them reconnection is impossible. In the long run we may be better off this way. If the special features of the developing brain persisted, the system might never become stable enough to maintain the long-term connections that we need to function consistently. The challenge faced by doctors and scientists today is not only to help injured cells recover the potential for growth that they enjoyed in the embryo, but to give them the right instructions for reconnection.

4 March 1989

Further Reading

In 'Wired for thought' (*New Scientist*, 28 August 1988) James Fawcett gives a concise account of brain development. For an elegant, readable and comprehensive treatment of the subject, see *Principles of Neural Development* by Dale Purves and Jeff Lichtman (Sinauer, 1985).

CHAPTER 22

The Nervous System: Repairs to the Network

Georgina Ferry

Children who suffer a prolonged shortage of oxygen at birth may be brain damaged. Young people who break a neck in skiing or motorcycling accidents may sever their spinal cord. Old people may fall victim to a stroke: a blocked artery or burst blood vessel in the brain, which kills brain tissue by depriving it of oxygen. All face the loss of abilities the rest of us take for granted, such as movement or speech. Tragically, in many cases, the loss is a permanent one.

Unlike other tissues, such as skin and bone, the **central nervous system (CNS)** – the brain and spinal cord – has very little capacity to repair itself. If a nerve cell is damaged in such a way that its **axon** – the long fibre that connects the cell to others in the CNS – is severed, the cell is unable to grow a new one. Without an axon, the cell is cut off from its partners in the network of cells that carry messages throughout the CNS. Nerve cells communicate with each other by electrical impulses that pass along the axon, the tip of which branches into hundreds of **nerve terminals**. At these terminals, the cells make contact with each other at junctions, or **synapses**. There is just a tiny gap between cells at the synapse, and one cell releases chemicals called **neurotransmitters** into the gap. The effect of these neurotransmitters is to increase or decrease the chances that the next cell in the network will act by firing an electrical impulse.

The axon is also important to the cell because it transports proteins and other essentials between the cell body and the

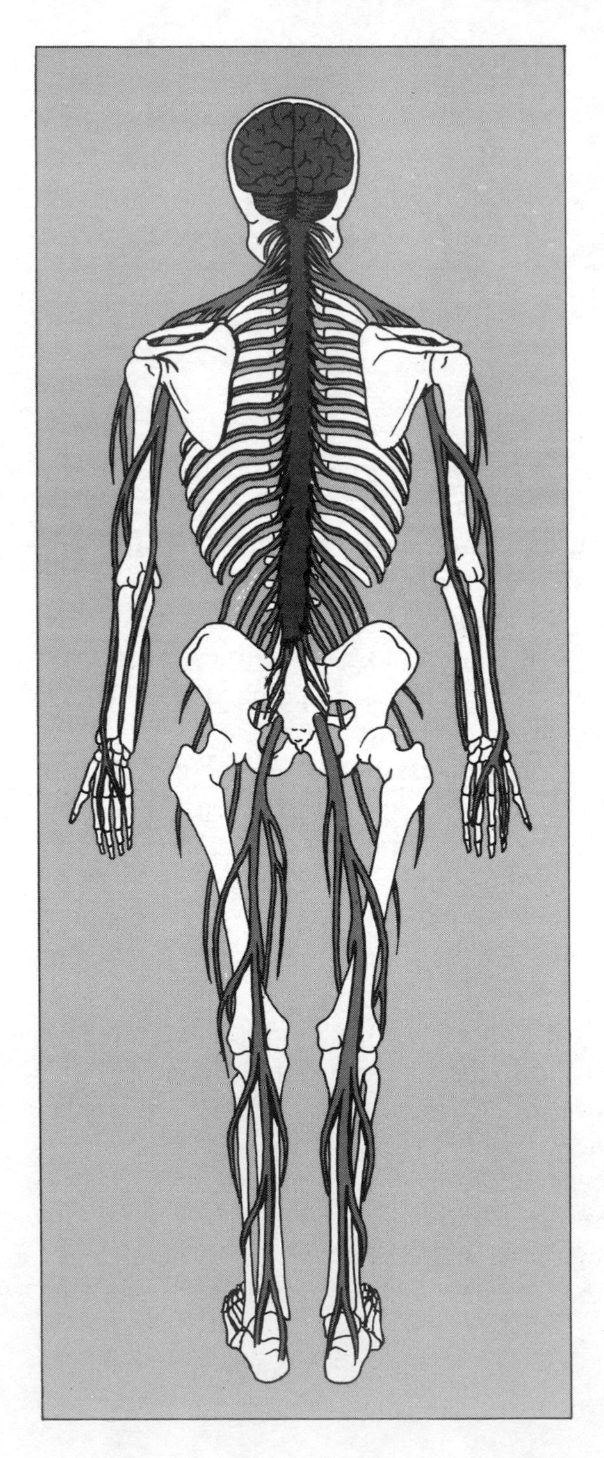

The central nervous system (dark grey) encompasses the brain and spinal cord. The peripheral nervous system (pale grey) extends from these to the limbs and organs

terminals. If the axon is severed these supplies are interrupted; as a result the whole cell **degenerates** and, ultimately, dies. That might not be the end of the damage. The cell will have been a vital link in a network of communicating cells; in some cases, after the loss of one such link, the rest of the cells in the chain begin to degenerate.

The nerve cells we are born with are already mature, and incapable of reproducing themselves by dividing. Those that die cannot be replaced. Surviving cells can take over the responsibilities of their immediate neighbours, but this response compensates for damage only over a very small area. For people with injuries to the brain or spinal cord, doctors can offer only limited treatments. After an accident, the first priority is to try to contain the extent of the damage. Once the patient is out of immediate danger, the long, slow task of rehabilitation begins. Where possible, doctors will teach patients skills that make use of the surviving networks of nerve cells in the brain to overcome some of their disabilities. The damaged areas themselves cannot be repaired. The goal of a great deal of current research is to understand why the cells of the central nervous system are so handicapped when it comes to making repairs. Understanding the mechanisms involved might lead to long-awaited treatments for brain damage.

For many years scientists debated whether the inability of brain cells to regenerate is part of their intrinsic genetic make-up, or whether the CNS is an unsympathetic or even hostile environment for a regenerating cell. In the past two decades it has become clear that **environment** is all-important in a cell's bid to achieve reconnection with other nerve cells.

Creatures high and low
Rewiring for beginners

This insight has partly come from studies of animals that have nerve cells that are able to **regenerate** quite happily. Lower vertebrates such as frogs and fish, for example, can carry out remarkable feats of rewiring. In the 1950s the Nobel prize-winner Roger Sperry carried out some classic experiments on regeneration in the CNS of frogs. He cut the optic nerve that connects a frog's eye to its brain, and observed that not only did the nerve fibres regrow, but they reconnected to appropriate sites in the brain. This reconnection was good enough for the frog to catch flies once more.

Simpler animals, such as the medicinal leech, are also adept at this kind of reconnection. The nervous system of the leech consists of groups of nerve cells, one for each segment of the body, linked in a long chain to coordinate its movements. John Nicholls, a British biologist working in Switzerland, found that even when the chain is cut right through, the axons reconnect so that the leech can swim again within a matter of weeks.

Scientists who study these lower animals are beginning to identify some of the conditions cells need to regrow and reconnect. Nerve cells from the leech, for example, can be removed and grown in a glass dish to determine their precise requirements. The cells, like those in any developing nervous system, need to attach themselves firmly to a surface, and only surfaces with the right kind of adhesive properties will do. They may also need a supply of specific chemicals called growth factors in the surrounding fluid. Further experiments are attempting to discover how cells recognize the partners with whom they must form connections. These studies have given scientists some idea of what to look for in understanding why cells in the central nervous system of higher animals fail to regenerate.

Central *v.* peripheral
What's the difference?

Big differences between animals with simple nervous systems and mammals might not seem that surprising. But in mammals there are equally dramatic differences between nerve cells within a single individual. Cells in the brain and spinal cord normally fail to regenerate, while those in the **peripheral nervous system** –

the network that connects the spinal cord to the limbs and organs – can regenerate fairly well, even in adults. If you break an arm or a leg, nerve cells are likely to be damaged, so that you lose some sensation or motor control in the affected part of the body. But if the damage is not too disruptive, or a skilled microsurgeon can reconnect the severed ends of the nerve, there is a fair chance that the cells will grow new axons. Unfortunately, they do not always reach the right destinations, especially if the target area is some distance away. In more successful cases, though, at least some of the lost functions will return.

How do peripheral fibres do this? At the beginning of the century, the Spanish neuroanatomist Santiago Ramon y Cajal suspected that the axons receive support and guidance from their surroundings as they regenerate. He also suggested that somehow this support is lacking in the central nervous system. The clearest indication of this comes from cells (such as sensory nerve cells) that have two fibres: one coming from outlying parts of the body bringing information, and another going into the spinal cord to transmit the information to the brain. The incoming nerve fibre can regenerate if it is damaged, demonstrating that the cell has all the machinery necessary for growth and reconnection. But if the other fibre is damaged, it grows only as far as the spinal cord; as it enters the cord it stops abruptly, and fails to make any connections.

What do peripheral nerves have that is denied to fibres in the CNS? An important feature of the environment of nerve cells is the supporting cells, called **glial cells**, that surround them. In the peripheral nervous system there is only one type: the **Schwann cell**. For each fibre, Schwann cells manufacture an insulating sheath of a fatty substance called **myelin**, which increases the speed with which an electrical impulse can travel down the fibre. Outside the myelin the Schwann cells lay down a **basement membrane** – a strong but flexible sheet of tissue that holds everything together. When a nerve is cut or crushed, the hundreds of thousands of nerve fibres that have been disconnected from their cell bodies gradually degenerate. But the outer membranes made by the Schwann cells remain, as do the

When a nerve fibre is damaged

Even when a nerve cell is fully grown it constantly manufactures new proteins in the cell body to maintain its structure. The axon, as well as conducting electrical impulses, has the job of transporting these and other essential supplies all the way to its terminals.

Peripheral nerve fibre

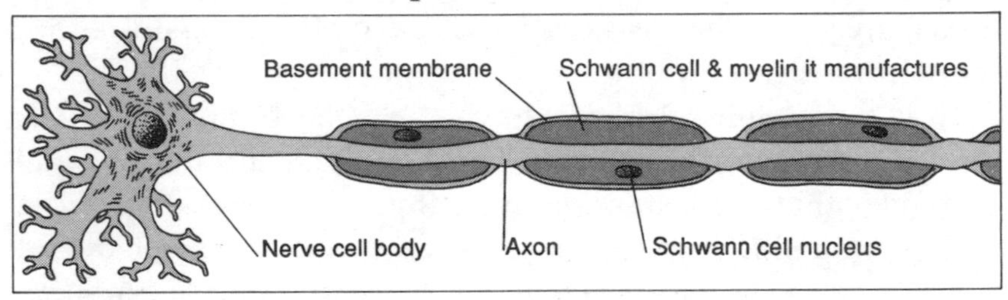

The normal peripheral nerve and its axon, surrounded by a sheath of myelin manufactured by the Schwann cells

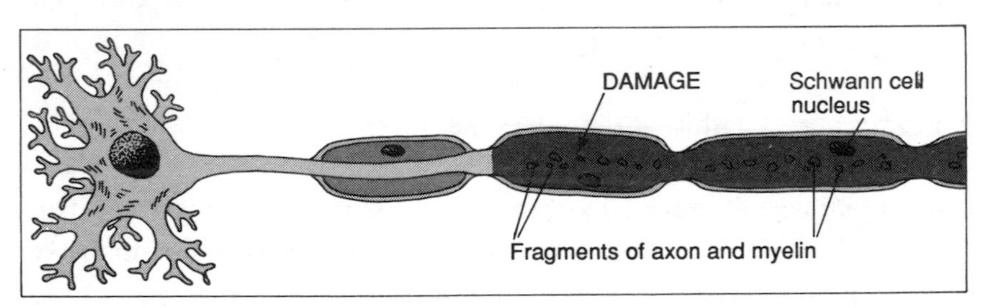

Beyond the site of damage (arrow), the axon and the myelin degenerate. But the Schwann cells and basement membrane survive, forming a continuous tube. The cell body undergoes many changes, such as increasing the synthesis of certain proteins, which may be important in helping the axon to regrow

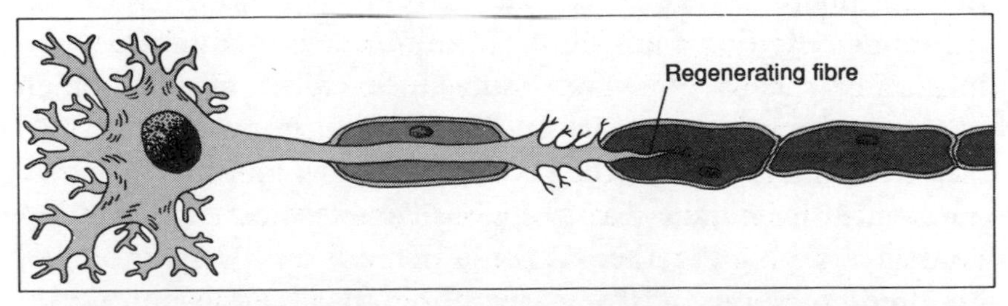

A growth cone develops on the tip of the axon stump, its branches seeking signals to guide its growth. When it locates the Schwann cell sheath, an axon grows along it

Central nerve fibre

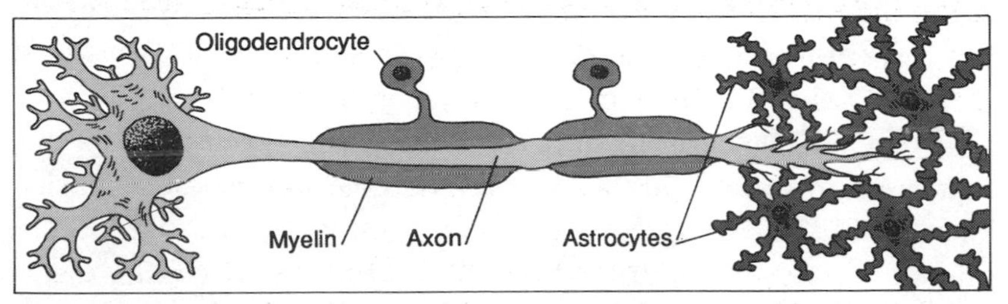

Similar changes may take place in a damaged nerve cell in the brain or spinal cord. But the oligodendrocytes also degenerate, and nothing is left to guide growing fibres. Astrocytes which congregate at the site of the damage may further increase the difficulty of the task. In addition, the myelin sheaths of undamaged fibres in the vicinity contain a protein that actively inhibits growth

The traffic is two-way, because the terminals also receive growth factors or other signals from their sites of contact, which in their turn have to be communicated to the cell body.

Cutting or crushing the axon brings all this to a halt, and every part of the cell reacts in its own way to the damage.

---

cells themselves. The stumps of fibres that are still attached to their cell bodies start to grow again in a very short space of time, possibly triggered into action by the sudden loss of contact with their targets. As long as these fibres are still inside the surviving Schwann cell sheath, or within very close range of it, they simply grow down inside it until they arrive back at the site, on a muscle or in the skin, where they terminated previously.

This pre-existing pathway is clearly a great advantage for fibres that may have many centimetres to travel. But Schwann cells offer growing peripheral nerve fibres more than just a conduit to guide them home. The basement membrane provides one of the 'stickiest', and therefore most attractive, surfaces there is for growing nerve fibres. In addition, they produce growth factors which diffuse out to the regenerating fibres and promote their growth. In an attempt to find out which of these was the most important to growing axons, scientists grew isolated peripheral nerve fibres in a glass dish with each component in turn. It turned out that no single component of the Schwann cell sheath could support their growth as well as the whole, living Schwann cells themselves.

Before the regenerating fibres can take advantage of all that the Schwann cells have to offer, they have to overcome an obstacle. The site of the damage to a nerve is usually a mess, with blood and decaying nerve fibres making it hard to reach the clear route through the sheath. However, white blood cells called macrophages converge on this debris and systematically clear it away. In the peripheral nervous system then, as long as the fibres can find their way into their sheath of Schwann cells, and as long as their path is not blocked, they will easily grow back and reconnect. Why does this not happen in the brain and spinal cord?

The CNS jungle
Hostile, or just unfriendly?

Damaged cells in the CNS do seem to make some attempt to regrow, but in normal circumstances they stop after a few micrometres. Unlike peripheral nerves they have no Schwann cells to guide and support them. Axons in the brain and spinal cord receive their insulating myelin from glial cells of another variety, the **oligodendrocytes**. These cells do not produce a tough outer membrane to keep everything in place while fibres regrow; they also tend to give up and die if the fibres they are associated with degenerate. So even if axons in the CNS did regenerate, they would be faced with a near-impossible task of navigating through a pathless jungle of nervous tissue to find their targets.

Meanwhile, another family of glial cells, the **astrocytes**, invade damaged areas in the CNS and form a tough, fibrous scar. In addition, there is the **blood-brain barrier**, a protective filter which prevents macrophages from entering the brain in large numbers, and so decaying material at the site of the

Brain defences

It might seem perverse that the most important organ in the body is one of the least able to cope with moderate amounts of damage. But the brain and spinal cord make up for their deficiencies in this respect by being equipped with special protection to minimize the chances of damage.

The **skull** and **spinal column** provide a bony armour that protects against all but the most severe of physical injuries. Inside this, the **cerebrospinal fluid** surrounds and protects the brain's delicate nervous tissue.

Then there are the blood vessels, which are very selective as to what they allow to pass through their walls. The cells that form these walls are tightly linked to one another so that only the smallest molecules can pass between them. They also contain mechanisms that control the rate at which molecules such as glucose can enter the brain.

This protective filter is called the **blood-brain barrier**. It ensures that the brain receives a steady supply of oxygen and glucose to fuel its intense activity, while keeping out many toxic substances and cushioning some of the effects of the sudden fluctuations in the chemical composition of the blood.

The existence of these security measures testifies to the enormous importance of the brain and spinal cord in the lives of human beings and other animals. But even so, in our high-speed, high-tech, occasionally violent and drug-dependent society, the defences are breached all too often.

---

damage is left lying around. Worse is to come. It now appears that oligodendrocytes and the myelin they make both contain a substance that actively inhibits regeneration.

A group of scientists led by Martin Schwab, working in Germany, and now Switzerland, is studying the circumstances in which nerve cells will grow, by looking at small groups of them in a laboratory dish. The team found that nerve cells taken from either the peripheral or central nervous system of new-born animals would grow into a piece of peripheral nerve.

But neither would grow into a piece of optic nerve, which resides in the central nervous system. By separating the components of the optic nerve, they discovered that oligodendrocytes and the myelin these glial cells make were effectively no-go areas for growing fibres.

Schwab and his colleagues have now succeeded in identifying a protein in the myelin that seems to be exerting this inhibitory effect. If they inactivate this protein, there is a marked change in the interaction between nerve fibres and optic nerve. The way they do this is to inject a **monoclonal antibody** (a molecule that attaches itself to a protein in such a way as to render it inactive) against this protein into the optic nerve. The axons, which previously avoided the optic nerve, now energetically invade it. In other words, once its inhibitory action has been neutralized, the optic nerve provides as good an environment for regeneration, in a glass dish, as the peripheral nerve.

With so much against it, it seems hardly surprising that the brain's attempts to repair itself will fail.

Nerve grafts
A bridge too far?

Peripheral nerve fibres can regrow, central ones cannot; the difference seems to lie in their environment, particularly the glial cells that associate with them. The experiment needed to clinch the argument seems obvious: will central fibres grow in a peripheral environment? For the past ten years Albert Aguayo and his colleagues in Montreal, Canada, have been addressing this question in rats – with spectacular results.

Aguayo's technique involves removing pieces of sciatic nerve (the main nerve to the leg, and therefore part of the peripheral nervous system) and using them as 'bridges' for damaged fibres in the CNS to grow along. Once the segment of sciatic nerve is removed from the rat's leg, the severed fibres degenerate; but the Schwann cells and their surrounding membranes remain, and even increase in number by dividing. Inserting one end of this piece of nerve into the spinal cord or brain has a dramatic effect on damaged nerve cells whose cell bodies lie nearby. Their axons begin to regenerate, entering the peripheral nerve graft and growing along its length. In some experiments Aguayo has seen these fibres grow for several centimetres, farther than they would normally grow in the animal. Where the bridge has connected two different regions of the central nervous system – the spinal cord and the brain stem, for example – axons have entered at each end and grown in both directions.

Aguayo's work has demonstrated once and for all that the cells of the central nervous system are perfectly capable of regenerating if they have access to the right environmental conditions. This would seem to suggest a route for treating people with damage to their brain or spinal cord. There is still

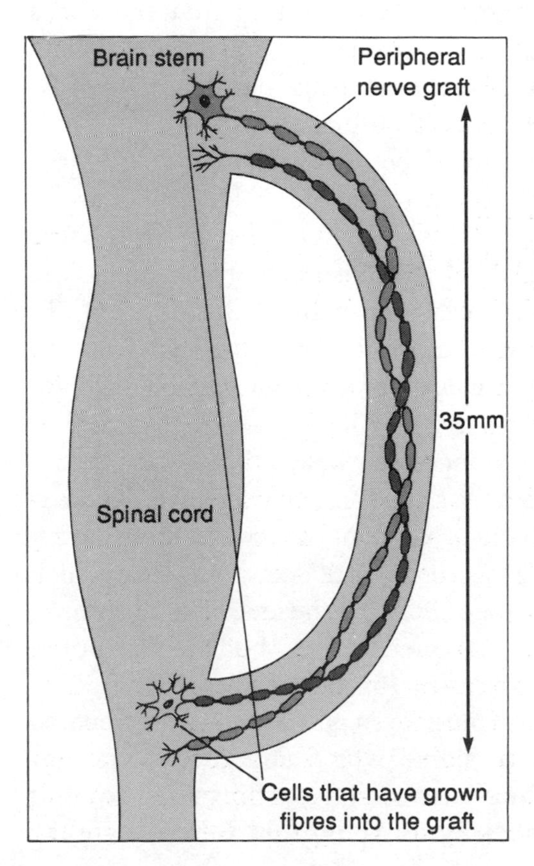

One end of a piece of peripheral nerve from a rat is inserted into the brain stem, the other into the spinal cord a few centimetres away. After several weeks, cells whose cell bodies lie near where the 'bridge' is inserted will send fibres through the peripheral graft

one serious obstacle to overcome, however. The fibres that grow so energetically through a peripheral nerve graft grind to a halt almost as soon as they re-enter the central nervous system. Not only will they not grow any more: so far there is no evidence that the axons can make connections with the appropriate cells. Bridges have been a remarkable demonstration of the capacities for growth of fibres in the CNS, but they have yet to restore the flow of useful information that is the essence of the nervous system's function.

Only connect
Beyond recognition

In the spinal cord, for example, it is essential that the nerve cells not only reconnect, but reconnect appropriately. The nervous system is arranged in numerous parallel networks, each carrying a different type of information – touch, pain, taste, vision and so on. It would be a disaster if the networks became crossed, so that the brain registered a signal indicating, say, a gentle touch as an intense pain. In the embryo these networks develop as a result of a complex interplay of factors ensuring that growing nerve cells reach the right 'address'. In the mature CNS this system seems to have broken down to some extent; even if fibres regenerate, they fail to recognize the cells they should be contacting.

There are, however, instances of new connections being formed in the mature brain. Where some axons have degenerated, disrupting their contacts or synapses with their target cells, surviving axons nearby will 'sprout' new branches which move in and occupy the space left by the decayed terminals. The branches grow only a few micrometres, but their response may go some way to compensating for damage nearby.

The phenomenon of sprouting suggests that the vacated synapses release a chemical signal which advertises that they are available and stimulates nearby nerve fibres to sprout. It also suggests that there is a way for the sprout to recognize the vacant site and make contact with it. Better understanding of these mechanisms might lead to a way of encouraging newly regenerated fibres to home in on appropriate sites of contact.

Why, in the course of evolution, should the CNS of animals and other higher vertebrates have lost the capacity to regenerate? It seems reasonable to suggest that it has something to do with the great increase in the complexity of the brain that underlies the superior mental abilities of these species. One possibility is that regeneration was abandoned to avoid an unacceptable risk of errors in rewiring. Perhaps more likely is the idea that mechanisms, designed to ensure that the multiple pathways of the brain develop according to the appropriate 'map', prevent further growth once that network is complete. We have begun to understand how some of these mechanisms operate. One day, we may be able to exploit our knowledge in developing new treatments for brain damage.

10 June 1989

Further Reading

This field of research is moving so fast that no one book can be completely up to date, and those available are written for the specialist. If you can get hold of it, read *Degeneration and Regeneration in the Nervous System* by Santiago Ramon y Cajal (Raoul May's translation, published in 1928 by Hafner Press). In this readable classic, Cajal raises many of the questions that scientists today are still trying to answer. Chapter 18 of *From Neuron to Brain* by S. Kuffler, J. Nicholls and R. Martin (second edition, Sinauer, 1984) gives a more up-to-date account. M. Chiquet's article for *Trends in Neurosciences* (January 1989, p.1) covers the latest discoveries on 'Nerve growth inhibition by CNS myelin proteins'.

CHAPTER 23

The Nervous System: Remaking the Brain

Georgina Ferry

Hundreds of people underwent operations during the 1980s to place cells in their brain which do not belong there. The aim is to reduce the symptoms of an incurable brain disease. Given the enormous complexity of the brain – billions of highly specialized cells, each with up to 1000 connections – does such an approach stand a chance of success?

A brain transplant, or more properly a **brain graft**, is a very different matter from other types of transplant. For a start, the surgeon is not transplanting a whole organ. Replacing a kidney is effectively a plumbing job, rather like replacing a filter in a domestic water supply. Brain grafting is more like electrical engineering. The brain consists of interconnected circuits that influence one another's activity. If one type of circuit is damaged, other connected circuits may no longer function properly, even if they themselves are undamaged. The principle behind grafting is to replace the faulty circuits, so that the whole system can function once more.

Humans suffer from various diseases that appear to damage the brain selectively, attacking a restricted area or population of cells. Among the most common are: Parkinson's disease, which affects muscle coordination, Alzheimer's disease, which causes memory failure in elderly people, and Huntington's chorea, which is hereditary and affects both movement and mental state in middle life. Strokes are not selective in the same way. Blocked or ruptured blood vessels may starve brain tissue of oxygen, so damaging a small area. In epilepsy, seizures begin with abnormal activity in a small area, which may spread to involve the whole brain. Neurosurgeons have suggested that all these diseases are possible candidates for treatment with grafts of healthy brain tissue. There are no drugs available at present that can offer a cure.

The idea that grafts might work in human patients stems from experiments on laboratory animals. These show that grafts of nervous tissue can not only survive in the brain of a laboratory animal but also make connections with some of the host's cells. The grafts can even correct behavioural problems arising from certain types of brain damage. This research has not only suggested a possible new treatment for human patients, but also increased our understanding of the brain's potential for growth and repair, and shed some light on the way its functions are organized. Elizabeth Dunn, an anatomist from Chicago University, demonstrated in 1917 that cells or pieces of nervous tissue could be successfully grafted from the brain of one newborn rat to another. But it was not until the 1970s that researchers in the US and Sweden improved on her rather hit-and-miss techniques, by painstakingly establishing the conditions necessary for grafted tissue to 'take' in another brain. In particular, this means ensuring that the new tissue has access to a supply of blood, while avoiding its rejection by the immune system. First, however, one has to choose appropriate tissue for grafting.

What to graft?
Cells on the grow

Traditional transplants involve surgeons taking the heart, a kidney or bone marrow, for example, from a living or dead donor, usually a mature adult. Things are very different with the brain. If the surgeon takes cells from a fully developed brain it disrupts their extensive connections. This is because the long fibres, or **axons** (see Chapter 21), which they use for communicating with other cells in the nervous system, are damaged. Once the brain cells are disconnected, nothing can be done to keep them alive; they certainly will not survive grafting to another brain.

Unlike other tissues in the body, the nerve cells in the brain are more or less mature at birth. Only during the development of the fetus do the nerve cells have the capacity to increase their numbers by dividing, to grow new axons over any distances and to make new connections. Most laboratory researchers in neural transplantation, therefore, concentrate on taking tissue from the brains of rat fetuses and grafting it into adult rats. Cells in these grafts are not yet fully mature and have not established their connections. As a result, they are still flexible enough to cope with transplantation to another host.

Brain cells are not all alike; they come in a variety of shapes and sizes according to the regions they occupy and the tasks they have to perform. Each, too, has a specific chemical identity. The cells of the nervous system communicate with one another by releasing a chemical, a **neurotransmitter**, from nerve endings at the tip of the axon. In its turn, the cell receiving the signal has specialized molecules called **receptors** on its surface (see Chapter 21). The receptors respond only to the appropriate neurotransmitter. The connection between cells takes place at the **synapse**, where one cell releases neurotransmitter across a tiny gap to another cell. Each cell has up to 1000 synapses. Now that neuroscientists have mapped some of this immensely complicated chemical switching system, they can choose cells for grafting according to the neurotransmitter that the cells contain. Just as with any precision instrument, you can repair the brain more effectively if you have the right parts for the job.

In a typical experiment, a graft consists of a few thousand cells, just a few cubic millimetres of tissue. This is inserted either as a solid piece, or as a suspension of **dissociated cells**. Dissociation involves treating the cells with an enzyme, trypsin, to loosen their grip on one another, and then shaking or stirring the tissue until it comes apart. The suspended cells are then drawn up into a syringe for injection. The size of the graft, and the method used to insert it, is one of the many factors

Rejection

The body's immune system provides an effective system of defence against invasion by 'foreign' organisms, such as bacteria and viruses. The same system normally ensures that any tissue not bearing the molecular label that identifies it as 'self' will automatically be attacked and destroyed (see Chapters 18 and 19). Patients receiving kidney or heart transplants have to be treated with drugs that suppress the immune system. The drugs protect the transplant, but leave the patient vulnerable to infection.

In the brain, things seem to be different. Brain tissue grafted from one strain of rat to another, or even from mice to rats, is rejected much more slowly than skin grafts. The brain is known as an **immunologically privileged site** for transplantation. But the privilege is only relative.

Grafts survive best in rats if they come from the brain of a rat that is genetically identical to the recipient. While grafts of brain tissue from strains of rat that are genetically different may not be rejected immediately, they may not survive more than a few months. Treating the host rats with an immunosuppressive drug ensures that the graft will survive.

Surgeons who have carried out brain grafts on humans disagree about the importance of using immunosuppressive drugs: some do, and some don't. Time will tell who is right.

---

that will determine whether it will 'take' satisfactorily in its new host.

The new home
Conditions for survival

For a graft to survive, its cells must quickly gain access to adequate amounts of oxygen and nutrients, normally by way of the host's own blood supply. Early attempts at grafting often failed because experimenters took a relatively large lump of cells and inserted it deep within the densely packed brain tissue. The cells at the centre of the graft were starved of oxygen and soon died. New blood vessels from the host grew into the graft, but too slowly to reach most of the tissue. Smaller chunks of tissue stand a greater chance of success, while dissociating the cells greatly increases the graft's chances of survival. In addition, injections of dissociated cells cause less damage to the host brain.

One way to improve the graft's chances of survival is to place it in one of the **ventricles**. These spaces within the brain are filled with cerebrospinal fluid, which appears to provide some nourishment for the graft before it develops its own blood supply. At the same time, the space in the ventricle allows plenty of room for growth. The graft will attach itself to the wall of the ventricle and begin to grow fibres into nearby brain tissue. While this method causes relatively little damage to the host brain, the graft can reach only those areas adjacent to the ventricle. Another option is to place the graft immediately below the cortex, where there is a bed of densely interwoven blood capillaries. New blood vessels quickly invade the graft, increasing the chances of the cells surviving. Again, the number of areas that can be reached by this technique is limited.

However, it is possible for neuroscientists to create similar conditions at sites elsewhere in the brain. Removing a small amount of tissue from an implant site ahead of the grafting operation enables a new network of blood vessels to build up around the walls of the artificial cavity. Researchers have found that a graft implanted a week or two after the cavity is created is more likely to survive than one implanted immediately after. One theory suggests that the damage leads to a build-up of special molecules called **growth factors**, which stimulate and nourish the growth of new tissue.

Successful grafting also depends on the age of the donor cells. The ideal cells are those that have reached the stage when they are in the process of growing axons. This takes place at different points in gestation according to the area of the brain

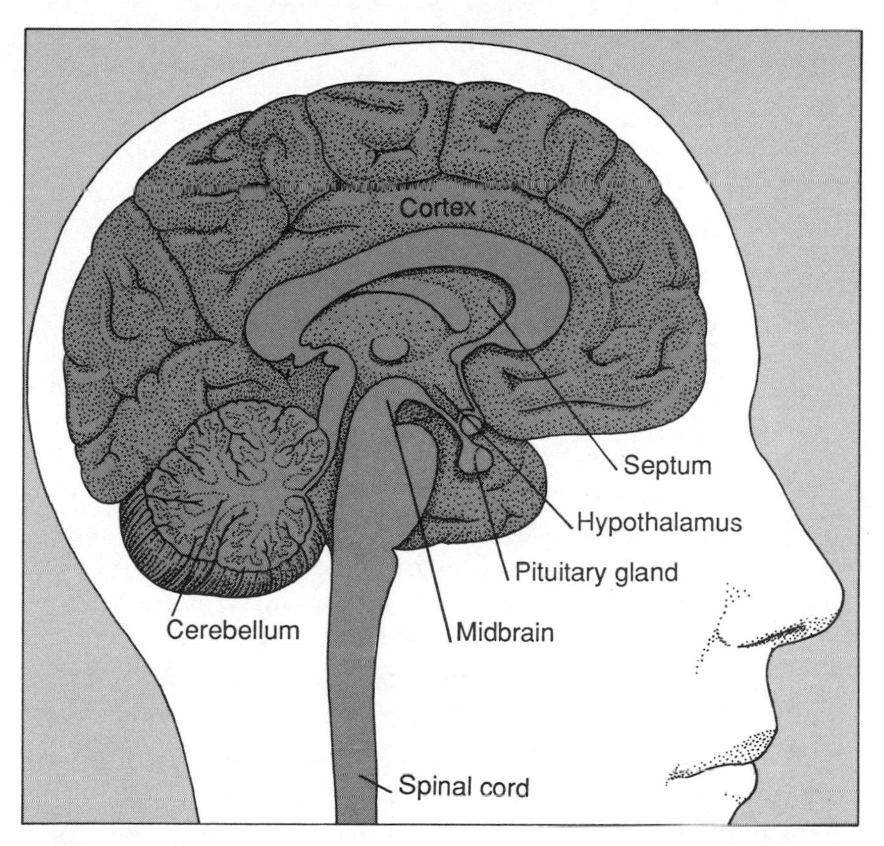

the cells come from. In rats, the best age of tissue for grafting varies from about 11 days' gestation to just before birth (21 days after conception) depending on the type of cell. Cells taken from a fetus that is more than 15 days old, however, do not survive well if they are dissociated. Doctors do not agree about the ideal age at which they should remove human fetal tissue for treating patients. Experiments, though, suggest that grafts taken from fetuses of between 9 and 12 weeks survive best.

Even under the best conditions, not all the cells in a graft will survive. In one experiment researchers estimated that only between one in 100 and one in 1000 grafted cells survived and grew. The graft, though, improved the rat's condition provided that more than 100 cells lived. In rats, therefore, one needs to inject at least 100 000 living cells to be sure of a good result.

How large a graft is needed to benefit a human patient is one of the many questions that have to be answered before doctors can consider the technique as a serious option in treatment.

What grafts do
Wiring or plumbing?

While the cells obviously have to survive if a graft is to play an active role in the host's brain, survival by itself is not enough. They have to establish some form of communication with other cells in the host brain. In the right circumstances, grafted cells will grow axons which penetrate short distances, making connections – the synapses – with cells in the surrounding tissue. The graft may also stimulate cells in the host to grow new axons into it and make connections there.

In the normal brain, communication between one region and another can take many different forms. Some circuits, or **pathways**, require transmission of precise information. One example is the visual pathway from the eye to the areas of the brain that make sense of what we see. In other regions, one area may simply act to **modulate** the activity of another, rather like a volume control. In grafts that are to play a modulating role, the availability of the appropriate neurotransmitter would be more important than restoring the circuitry precisely. Normal movement, for example, depends on connections between the substantia nigra and the striatum, an important motor control centre. The cells from the substantia nigra send fibres to the striatum, where they release the neurotransmitter **dopamine**. In a rat, an effect of damage to the nigro-striatal pathway on one side of the brain is to cause the animal to turn in circles.

A graft of cells from the substantia nigra of a fetus can reduce the rat's tendency to gyrate. Researchers place the graft, not at the site of the host's substantia nigra, but directly into the striatum. There it makes some connections with the host

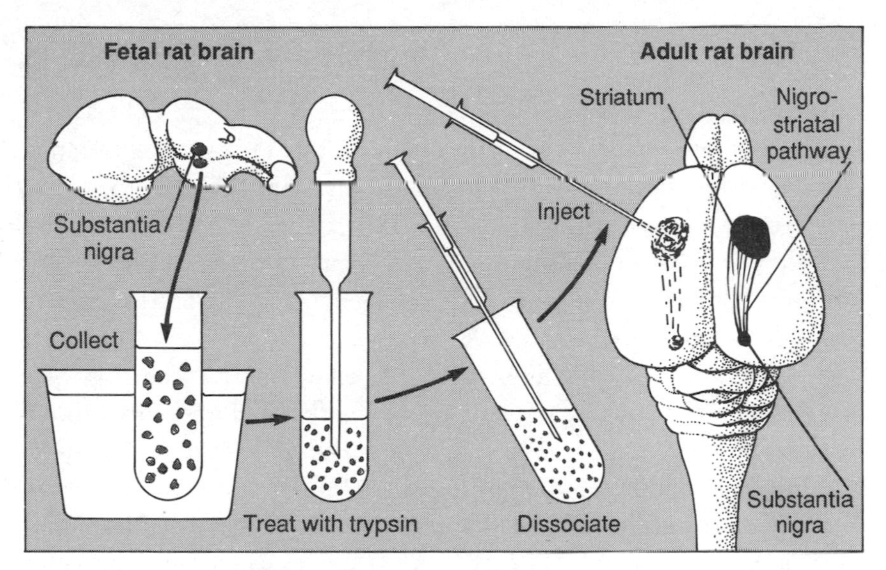

An adult rat with a damaged nigro-striatal pathway will turn in circles. Cells from a fetal substantia nigra implanted into the striatum will once more set the rat on a straight course

cells, where it releases dopamine. The graft enables the striatum once more to keep the rat on an even course. This is despite the abnormal position of the graft, and the fact that it lacks many of the connections of a normal substantia nigra. (This work has important implications for humans with Parkinson's disease. Cells in their substantia nigra have died, leading to a lack of dopamine in the striatum.) It is worth noting that this type of graft will form connections only if the host brain is missing its normal source of the transmitter. Receptor sites that are left empty can then be reoccupied by axons growing out from the graft. Animals gradually lose many of their brain cells as they age; in rats, at least, grafts can compensate for some of this loss.

Like elderly humans, or patients with Alzheimer's disease, old rats have a poor memory. The hippocampus plays an important role in storing new memories, receiving input from the septum via a pathway which uses **acetylcholine** as the neurotransmitter. The concentration of acetylcholine in the brain tends to decline

Human cells for human brains: a new cure?

The cells in a brain graft in laboratory animals survive. They can also make connections with the host brain, and produce the neurotransmitter for communication. On the basis of these experimental findings, doctors have made a number of attempts to use grafts to treat human patients with brain damage.

Most of the patients who have received brain grafts suffer from **Parkinson's disease**. This disease results from the death of cells in the substantia nigra, a small area within the midbrain. In a healthy brain, these cells communicate via a long pathway with the caudate and the putamen, collectively known as the striatum, which are involved in controlling movement. Here nigral cells release the neurotransmitter dopamine. The death of the substantia nigra cuts off the supply of dopamine to the caudate and putamen.

This has profound effects on people's ability to control their muscles voluntarily. Patients are not paralysed, at least not in the early stages of the disease, but they may have difficulty beginning a movement. Their limbs may also tremble uncontrollably or become rigid, and they may walk stooped forward, with rapid shuffling steps. The disease tends to strike in middle to old age. The symptoms worsen progressively over time, and there is no cure at present.

A drug, L-dopa, can raise the levels of dopamine throughout the brain, and most patients find it helpful. With time, though, the drug tends to become less effective, and it can have distressing side effects. The promising results from animal experiments, using grafts of fetal brain tissue, have encouraged the idea that a graft of dopamine-producing cells might benefit people with Parkinson's disease.

Because of the ethical issues involved in taking grafts from the brains of aborted human fetuses, at first doctors used a different source of tissue. The cells of the **adrenal medulla**, the inner part of the adrenal gland, produce various chemicals including dopamine. Experiments with rats had suggested that these cells could also work as dopamine-producing grafts. One advantage of using tissue from the adrenal medulla is that the graft can be taken from the patient's own adrenal glands. This avoids the possibility of the graft being rejected.

Swedish doctors attempted such grafts in the early 1980s, but they were largely unsuccessful. Then in 1986 Ignacio Madrazo, a surgeon in Mexico, reported excellent results in eight patients, prompting a wave of attempts to repeat these findings, mostly in the US. To date, more than 300 such operations have been performed. The results suggest that about 30 per cent of the patients have improved slightly, while others have died as a result of the operation. There is, however, little evidence that the grafts survive.

Scientists working with laboratory animals have always obtained much better results with grafts of fetal brain tissues. In Sweden, doctors worked with lawyers, philosophers and others to draw up ethical guidelines for using brain tissue from aborted fetuses to treat patients. Olle Lindvale, Anders Björklund and their colleagues at Lund carried out their first operations in autumn 1987 with grafts of dopamine-producing tissues from fetal brains. Doctors in other countries, including Britain, Cuba, China, Mexico and the US, have also begun, on an experimental basis, to treat Parkinson's patients with human fetal grafts.

Does the treatment work? The answer is still unclear. The patients who have received grafts tend to be those in whom the disease is far advanced, who may not be the best candidates for the technique. The initial results on the first two Swedish patients were disappointing, but the condition of one of them has since improved a little. Other doctors have claimed greater success, but with so few patients it is difficult to evaluate their claims. In none of the patients can grafts be said to constitute a cure. The patients

A graft for Parkinson's disease

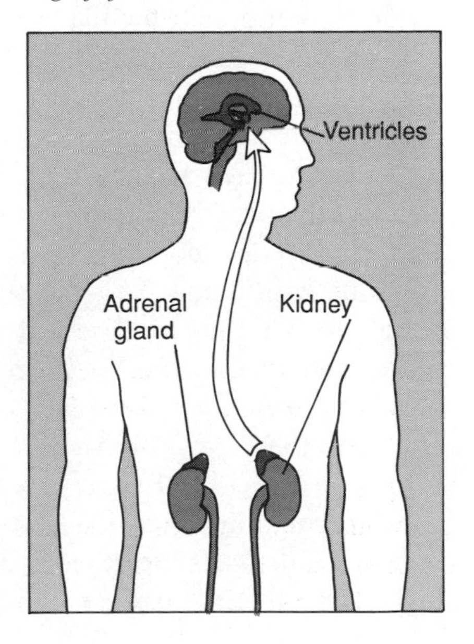

still have Parkinson's disease, although their symptoms may, in some instances, be less severe than before.

The technique is so new that many questions remain to be answered. For instance, do the dopamine-producing cells in the graft themselves fall victim to the disease? How long will the grafts survive? What steps, if any, need to be taken to prevent rejection of the foreign fetal tissue?

Over the next few years, the answers will determine whether brain grafting takes its place in the armoury of technological approaches to disease.

---

with age. Forgetful old rats given a graft of cells from a fetal septum can once more learn, and remember their way around mazes. Again, the precise circuitry is not important as long as the acetylcholine reaches the right target cells.

Some brain regions act as chemical switches, turning on the activity of another area. For example, the hypothalamus releases a hormone which in turn stimulates the nearby pituitary gland to produce more hormones. These act on the ovaries and testes, ensuring normal sexual development. There is a strain of laboratory mouse which is genetically unable to produce the hypothalamic hormone, so neither males nor females mature sexually. A graft of fetal hypothalamic tissue from a normal mouse sets the hormone cascade into action; females with grafts have mated and given birth.

Recently, experiments with another type of graft have shown that it may be possible to rebuild circuits more precisely. Animals with damage to the striatum, which regulates movement, are overactive. They also have difficulty carrying out tasks such as retrieving food accurately with one or other paw. Grafts of striatal tissue can reduce these problems. To be effective, these grafts have to be placed in the striatum itself, not, as in the case of substantia nigra grafts, in the areas with which it communicates.

If striatal grafts work, it must be because axons from the nigro-striatal and, possibly, other pathways are making contact

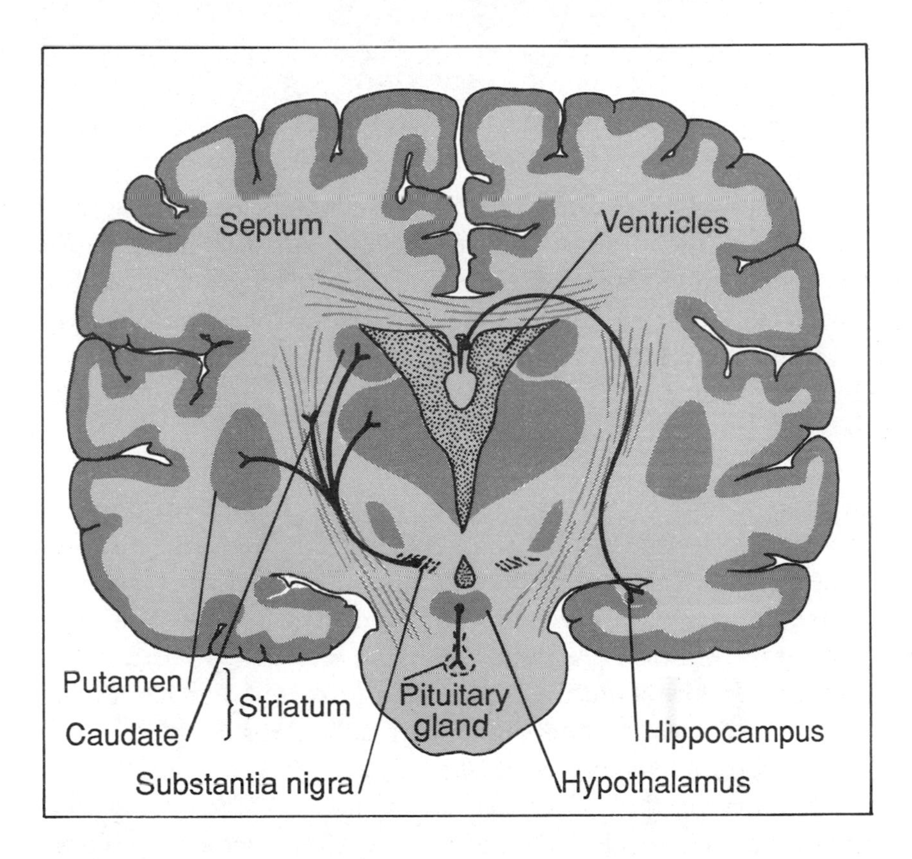

Three modulatory circuits, or pathways, in a human brain. These pathways rely on neurotransmitters for communication. In experiments with rats and mice, grafts of cells that produce the appropriate neurotransmitters have successfully corrected problems arising from damaged cells or faulty pathways. This cross-section of the human brain is idealized. In reality, the pituitary gland and striatum would not be visible: the pituitary gland is farther forward, and the striatum closer to the back of the brain

with them and influencing their activity. In addition, the grafts' effectiveness should depend on their sending axons out into other areas and making connections there. Researchers who have studied the animals' brains under the microscope report that there are axons running to and from established striatal grafts. Their pattern suggests that they are indeed re-establishing the circuit that was broken when the striatum was damaged.

New directions
Future of brain grafts

The results of these and other experiments suggest that grafts could hold the promise of new treatments for humans. In particular, grafts might repair pathways that have a modulatory role. Neurosurgeons have yet to prove conclusively the effectiveness of brain grafts in humans, however, and many practical questions remain to be resolved. In the meantime, other research is spawning new questions of its own.

The techniques of genetic engineering mean that, in theory, it is possible to tailor any cell in the body to fit different requirements. For example, one could take a skin cell and alter its genes so that it would make dopamine. Such cells will multiply happily in a laboratory dish, and could then be used as a graft to treat Parkinson's disease. This would have many advantages. Patients could donate their own skin cells, removing both the problem of rejection and the ethical difficulties of using tissue from fetuses.

Many people find it hard to accept the idea of tinkering with the brain, which still holds many associations with the mind or personality. Such views may be out of date now that we are beginning to understand the workings of the brain well enough to consider repairing it. As for Dr Frankenstein's scheme of transplanting a whole brain to a new body – that remains in the realms of science fiction, where it belongs.

11 November 1989

Further Reading

Articles in *New Scientist* on brain grafts include 'Brain transplants: fiction and reality' (20 September 1984) and 'New cells for old brains' (24 March 1988). A paper 'Mechanisms of action of intracerebral neural implants' by Anders Björklund and his colleagues (in Volume 10, p.509 of *Trends in Neurosciences*) details the ways in which grafts can influence the brain's activity.

CHAPTER 24

The Body's Protein Weapons

Frances Balkwill

For centuries, scientists have attempted to understand how the body coordinates its fight against disease, in the hope that this will lead to new therapies. Much is now known (see Chapters 18 and 19). Among the more recent discoveries are **cytokines**.

Cytokines have a variety of other names – lymphokines, monokines and interleukins. They are proteins that act as chemical messengers in the human body. Cytokines are made by many different cells, especially the cells of the immune system, and are usually produced when and where they are needed. These 'protective' proteins help to defend the body against attack by viruses, bacteria and fungi, and they also repair damage. They do this by signalling cells to change their behaviour in a number of ways: to start and stop growing, for instance, or to engulf invading bacteria more rapidly and efficiently. Some cytokines encourage the cells of the blood-vessel walls to stick to blood cells and let these blood cells get through the vessel into the tissue. Some can protect cells from viral infections by altering them so that the virus is unable to take over and produce more viruses.

Cytokines can deliver a number of different messages to cells, or they can bring the same message. The way this message is interpreted depends on the cytokine, which cell it acts on and what other messages that particular cell is receiving at the same time. Cell biologists have discovered at least 15 different cytokines so far. But they do not know why there are so many cytokines delivering so many messages, or why each cytokine doesn't just deliver one simple message. This complex and subtle system means that the body responds quickly and efficiently to infection and damage, but can shut down that response quickly.

Cytokines deliver their messages by sticking, or binding, to other proteins called receptors on the surface of cells. Different cytokines bind to different receptors. The structure of these receptors is such that the cytokine binds to that particular protein on the surface of the cell and to no other, rather like a key fitting into a lock. This action triggers off other signals that go to the nucleus very rapidly. These second signals change the proteins that the cell makes, and this alters the behaviour of the cell.

Scientists have known about cytokines and their potential for a long time but in the past they were very difficult to purify. Cytokines are extremely potent. For instance, one million millionth (10^{-12}) of a gram of one cytokine, alpha interferon, is

Cytokines interact in a complex network to control immune defence cells when they respond to 'foreign' and potentially harmful bacteria (antigens)

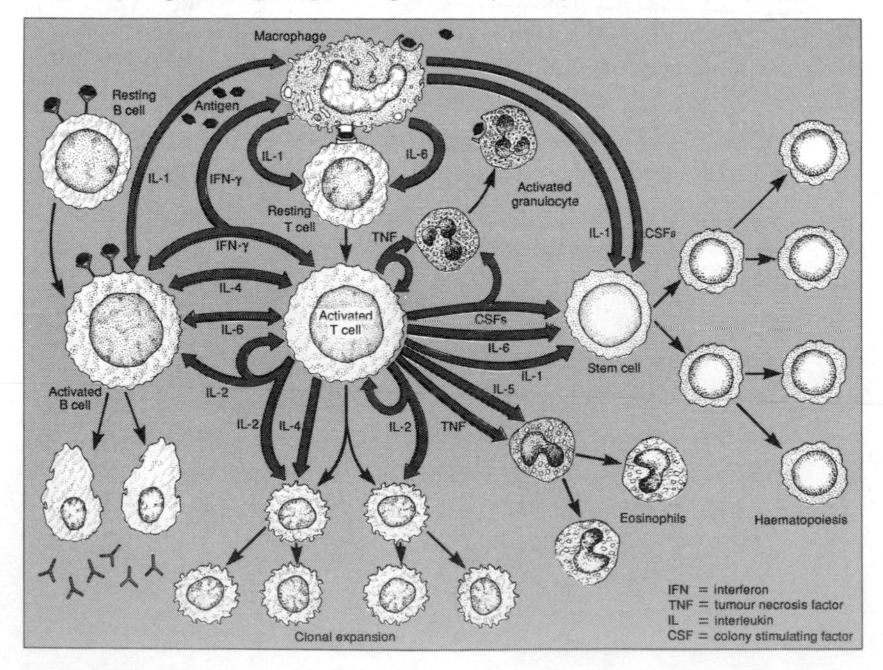

sufficient to protect one million cells in a test tube from attack by 10 million virus particles. In the past, the only way scientists could make a cytokine was to take some cells out of the body, grow them in a test tube and add a virus, bacterium or other harmful agent. After a few hours the liquid in which the cells grew would contain a range of cytokines, but for every cytokine protein there would be hundreds of thousands of other irrelevant molecules. It was very difficult to purify a single cytokine.

Once geneticists had developed recombinant DNA technology, researchers were able to purify the 'blueprint' of the human cytokine proteins and make bacteria produce it in vast quantities. Bacteria grow in very simple liquids, and purifying the human cytokines produced by bacteria is a very simple and economic process. So, at last, scientists have enough of these powerful messengers, the body's own weapons, to use in the fight against disease.

There are probably more cytokines still to be discovered, but those so far identified include the interferons, tumour necrosis factor, lymphotoxin, interleukins 1, 2, 3, 4, 5 and 6 and the so-called 'colony stimulating factors'.

Cytokine family: Interferons

In 1957 two scientists working in London isolated a protein that would protect other cells from attack by a wide range of viruses. Alick Isaacs and John Lindenmann called this protein, which had no effect on the viruses themselves but somehow changed the cell, the **interferon**. They had named the first cytokine.

The initial interpretation of important scientific discoveries is often deceptively simple and interferon was no exception. Isaacs and Lindenmann did not realize that there were many interferons and that some of their properties would be shared by the other protein molecules that defend the body. Scientists

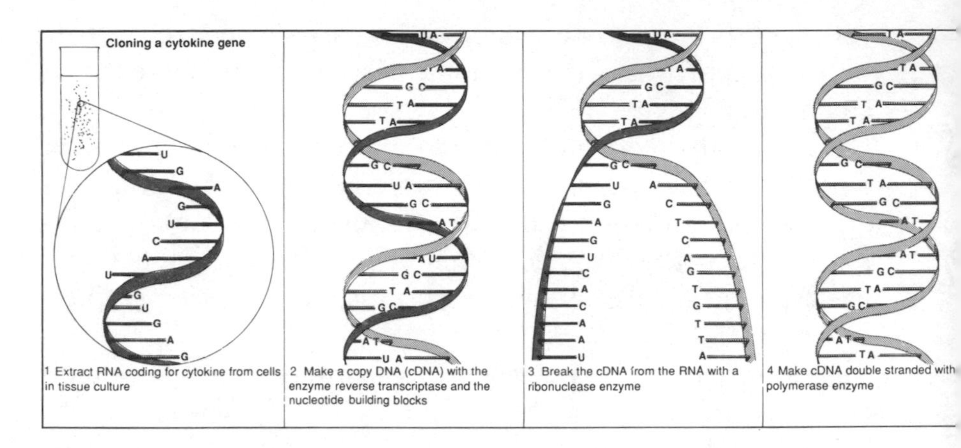

1 Extract RNA coding for cytokine from cells in tissue culture

2 Make a copy DNA (cDNA) with the enzyme reverse transcriptase and the nucleotide building blocks

3 Break the cDNA from the RNA with a ribonuclease enzyme

4 Make cDNA double stranded with polymerase enzyme

now know that each species of vertebrate possesses several different interferons (human beings have three types: **alpha**, **beta** and **gamma**).

There are 15 or more different alpha interferons but only one beta or gamma. Alpha and beta interferons bind to one receptor, and gamma to another on the cell surface. That was not the only complexity. Although the preparations of interferon were very impure, medical researchers soon realized that interferons did more than protect cells from viral infection. For instance, they could stop cells growing and they could stimulate the cells to produce more of the **MHC proteins** that are important to the human immune system (see Chapter 19). The researchers also found that these impure preparations of interferon could slow down the growth of cancers in animals.

For the past eight years interferons have been used in the treatment of many diseases. Clinicians have successfully treated some cancers of the blood cells with alpha interferon. In chronic myelogenous leukaemia, for instance, interferons act slowly. Over several months the number of white cells in the blood begins to fall until it reaches normal levels. Even more interestingly, over subsequent months, or even years, the number of cancer cells in the bone marrow begins to fall in some patients. No other drug can produce this result in the bone marrow. Treatment with alpha interferon will also shrink tumours (lymphomas) of the lymph nodes and reduce tumours and prolong life in a small minority of patients with the skin

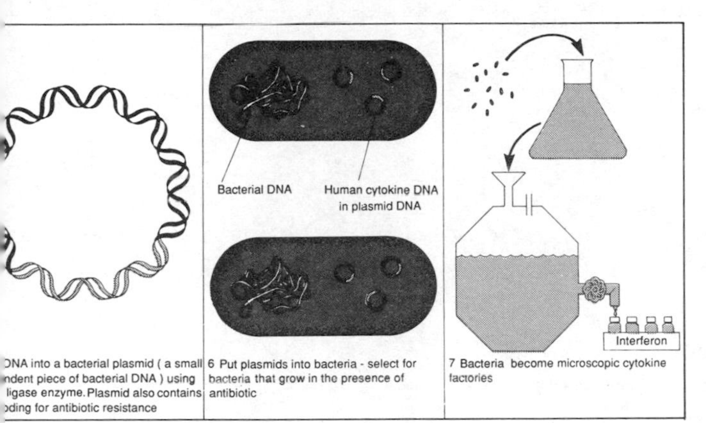

In 1977 the world supply of alpha interferon was less than one tenth of a gram. It took a year to make this from blood cells infected with viruses. Today, this amount can be produced in a day by bacteria containing the cloned gene, using a small fermentation tank

cancer melanoma. Unfortunately, there is no way of identifying those few patients with melanoma who will respond. So far, alpha interferons do not seem to work on the common cancers in the West – of the breast, bowel and lung.

The evidence suggests that alpha interferons act directly on tumours, inhibiting the growth of the cancer cells and making them behave more normally. Although trials are at an early stage, the two other types of interferon, beta and gamma, seem to work against the same spectrum of cancers.

In the early 1960s scientists hailed interferons as the antibiotics of viral infection because they were active against such a wide range of viruses. However, interferons do not act on the viruses but on the cells they infect. In experiments on animals, interferons are very effective in preventing viral infection if they are given before the virus, but they do not work once the infection is established. There are, however, important chronic viral infections that alpha interferon can help. Some patients who become infected with acute hepatitis B go on to become carriers of the virus and suffer from chronic liver disease. If they are treated with alpha interferon, the levels of various viral proteins in their blood reduce. What is more important, the number of infectious viruses drops, and the appearance and function of the patient's liver improve.

Genital warts are caused by a papilloma virus and are sexually transmitted. Treatment with alpha interferon, particularly when it is given directly into the warts, is highly

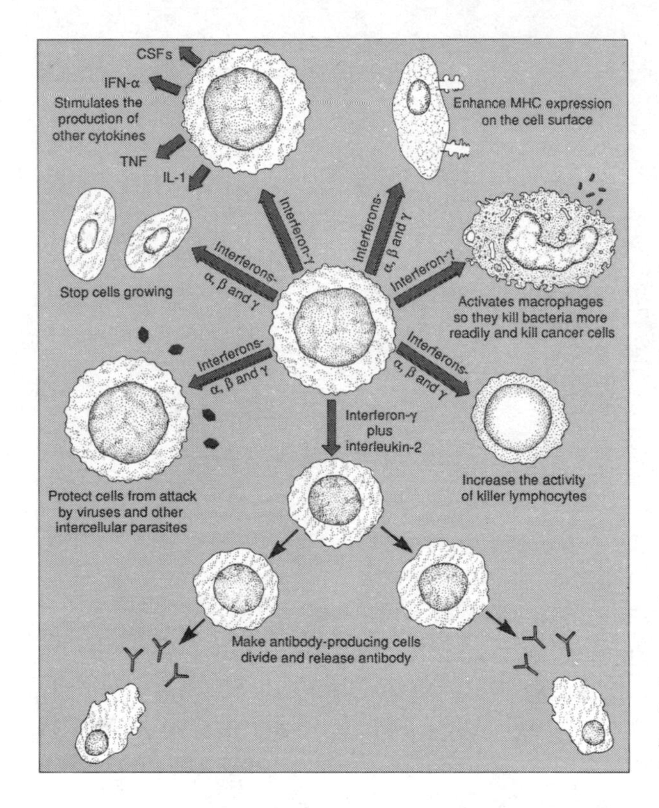

Interferons are typical cytokines. They change the behaviour of cells and make them produce other cytokines

effective. A nose spray containing alpha interferon helps to prevent colds developing in families where one member is already infected. However, treatment can only be given for six days because after this the treatment causes nasal stuffiness and nosebleeds and could be worse than getting a cold. Gamma interferon has been used with some success in patients with one form of leprosy, and may be useful in other parasitic infections.

Cytokine family: Tumour necrosis factor

Clinicians have known for several hundred years about one of the activities of another cytokine, **tumour necrosis factor**. (Necrosis means the death of living tissue.) In a few cancer patients with severe infections, tumours shrink and even disap-

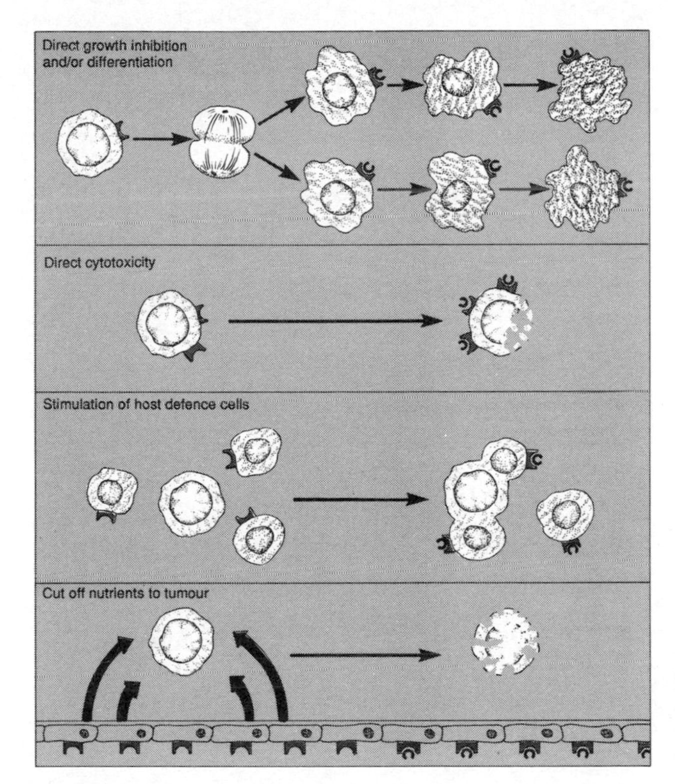

Cytokines (C) may stop cancer cells growing in a number of different ways

pear. At the turn of the century, researchers treated the cancer with a crude bacterial vaccine in an attempt to produce evidence to back up these observations. While they recorded some successes, it was 50 years before they discovered that this was due, at least partly, to the body's production of two cytokines, tumour necrosis factor and **lymphotoxin**, in response to the infection.

Cloning tumour necrosis factor and lymphotoxin raised a great deal of excitement. Were they really the body's own weapons against cancer? As quantities of these cytokines began to be passed around the world's laboratories, scientists discovered new properties. For instance, although tumour necrosis factor killed some cancer cells, it could also stimulate the growth of connective tissue cells. It had very powerful effects on the cells of blood-vessel walls, making them more 'sticky' to blood cells and helping blood to clot. Tumour necrosis factor could protect some cells against viral infection – but it could not protect as many cell types as the interferons. Unfortunately, this cytokine has proved disappointing in trials so far. Although several hundred cancer patients have been treated worldwide, few have responded, except when tumour necrosis factor is injected directly into the tumour, something that is not practical in most cancers.

Cytokine family: Interleukins

Scientists discovered other cytokines on several different occasions, but once they had purified the proteins they found them to be identical. They isolated **interleukin-1**, for instance, in a number of contexts: as a fever-producing factor (an endogenous pyrogen), as a stimulator of bone cells, as a factor which induces the breakdown of cartilage (catabolin), and as a factor involved in the growth and differentiation of white blood cells (haemopoietin).

The cytokine **interleukin-2** acts mainly on white blood cells called **lymphocytes**, making them grow and divide more rapidly in tissue culture. One of its most interesting properties is its ability to turn these lymphocytes into highly active 'killer' cells, which recognize and destroy cancer cells or damaged cells. A series of experiments on animals carried out in the US in the early 1980s showed that if such **lymphokine-activated killer (LAK) cells** were generated outside the body by incubating them with interleukin-2 and then given back to animals along with more of this cytokine, lung tumours would become smaller and the life span of animals would increase.

Many clinical trials are now under way to evaluate this treatment of cancer. Initially, vast quantities of the patient's white blood cells were removed using a machine not unlike those used for kidney dialysis. The cells were incubated with interleukin-2 for three to four days and then returned to the patient with more interleukin-2. More recent approaches try to produce LAK cells in the body by injecting the patients with interleukin-2 alone. Researchers are also trying different doses and methods of giving it to minimize side effects, which can be very severe. Whatever the type of treatment, there is no doubt that interleukin-2, alone, or in combination with these LAK cells, attacks cancer, and can make tumours shrink and occasionally disappear in about 20 per cent of patients. It is particularly effective in melanoma, kidney cancer and lymphoma. There is, however, much more experimental and clinical work to be done before it is clear whether interleukin-2 has any place in the future treatment of cancer.

One way cytokines may help to fight infectious disease is by improving the body's response to vaccination. Because interleukin-2 stimulates lymphocytes, this cytokine may be particularly effective. In studies on animals, interleukin-2 improved the potency of vaccine against haemophilus pleuropneumonia in pigs, a disease which is becoming widespread. Experiments are also now under way to see if interleukin-2 can improve the potency of the rabies vaccine. Interleukin-2 may also help some patients who have deficient immune systems.

Another cytokine, **interleukin-5**, seems very specific, at least in human beings, as it acts only on a very specialized form of white blood cell, the **eosinophil**, which is important in allergic responses and resistance to parasites.

Cytokine family: Colony stimulating factors

One group of cytokines is called the **colony stimulating factors** because cell biologists first identified them by their ability to make blood cells grow in clumps, or colonies, under laboratory conditions. These cytokines may be useful to patients who have too few of these cells, which are important in fighting infection.

Infection is the most common cause of death in cancer patients being treated with drugs (chemotherapy) because this treatment is highly toxic to the bone marrow. During treatment the number of white blood cells in patients' blood can fall to dangerously low levels and they may succumb to viruses and bacteria. Clinicians have given colony stimulating factors to patients after they have received chemotherapy. The bone marrow then recovers much more quickly from the toxicity of the cancer therapy, the number of white cells in the blood of patients does not fall so dangerously low and they do not get infections. Some people believe that more cancers could be cured if higher doses of chemotherapy could be given, but such doses would completely wipe out the patient's immune defences. The addition of colony stimulating factor to the current therapies removes this worrying aspect of toxicity.

These cytokines also stimulate recovery of the bone marrow in patients who have received large doses of irradiation to kill tumour cells in their bone marrow. It also helps to boost the white cells in patients with AIDS. One feature of the network of cytokines is that one often enhances the action of the others. Once the interactions between cytokines are more fully understood, combinations may be more useful than single cytokines.

Cytokine family: The darker side

These potent chemical messages can also harm us. If the cell is altered so that it produces the wrong cytokine, or too much cytokine for too long, or if it misinterprets the cytokine messages, this can cause problems. It now looks as if some of the symptoms of human diseases are due to overproduction of cytokines in the body. Paradoxically, switching off cytokine messages or inhibiting their actions may actually alleviate the symptoms of some diseases.

The first clue to this came when interferons were given to human patients. These natural proteins produced fevers, aches, shivering, drowsiness, fatigue and loss of appetite. This flu-like syndrome led to suggestions that naturally produced interferon makes people feel ill during an infection. Biologists now know, however, that other cytokines, particularly interleukin-2 and tumour necrosis factor, also cause high fevers, and all cytokine treatments, with the exception of the colony stimulating factors, produce quite unpleasant and similar side effects.

Another symptom of disease is loss of weight. Scientists in New York identified a protein which was produced by rabbits with a parasitic infection that interfered with the ability of the rabbits to use fats from their diet. When researchers purified this protein they found it was identical to tumour necrosis factor. This defence protein, produced to fight the infection, was contributing to the symptoms of the disease. In acute infections that can be fatal, such as meningitis, patients go into a state of shock. This shock quickly shows itself in the body through effects such as a drop in blood pressure, blood clots in many organs, fever, diarrhoea and organ failure. There is strong evidence that too much tumour necrosis factor contributes to this. Giving this cytokine to rats at a high enough dose can reproduce the signs and symptoms of a massive infection.

Treatment with toxins

In 1891 a young surgeon called William Coley who lived in New York was greatly distressed by the rapid death of a young patient with a bone tumour. Looking back over case histories he came across a patient with a recurring neck tumour who had a severe attack of erysipelas, a common bacterial infection of the skin in those days. His tumour completely disappeared, and he was cured.

Coley decided to treat some patients with filtered and inactivated liquid from bacterial cultures.

More than 1000 patients were treated with 'Coley's Mixed Toxins' in the next 40 years and, in 270 of these, tumours became necrotic (dead) then shrank. Some, where case histories are carefully detailed, were undoubtedly cured. But Coley's toxins fell out of fashion. Although Coley was often successful, other doctors did not follow his instructions carefully enough. He insisted that a high fever should be caused with each injection and that therapy be given for a long time. Also, the standard of commercial preparation of the toxins was very variable, and 'modern' techniques for treating cancer, such as drugs and irradiation, were being developed.

Coley's toxins were still studied in laboratories, however, and researchers found that the active ingredient was a highly toxic constituent of bacterial cell walls called endotoxin. It took 40 years for scientists to find that it was not the endotoxin but cytokines produced by cells in the animals treated with endotoxin that made tumours shrink. It was another 10 years before the tumour necrosis factor gene was cloned and a pure cytokine made available.

Was this really the body's own weapon against cancer? Modern trials with pure tumour necrosis factor have not cured any cancer patients. Why? Did Coley's toxins induce a cocktail of cytokines or is there another more active host factor yet to be discovered?

The researchers in New York also found that if antibodies specially made to fight tumour necrosis factor were given to animals before a potentially lethal dose of bacteria, they 'mopped' up this cytokine and the animals did not die. Moreover, clinicians found tumour necrosis factor in the serum of patients ill with meningitis and who went on to die, but not in those who recovered. Several drug companies and research institutes are now working on ways of neutralizing tumour necrosis factor, particularly with antibodies.

The cytokines that stimulate cell growth may actually be involved in the process of cancer. One theory of cancer suggests that cells become malignant because they produce a growth message that instructs them to divide repeatedly. This is called the autocrine hypothesis. Japanese scientists recently showed that the cytokine **interleukin-6**, which is a potent stimulator of the growth of antibody-producing B cells, is produced by cancerous antibody cells (**myeloma cells**) in human beings. Antibodies which stop interleukin-6 acting will stop myeloma cells growing in the laboratory. However, no one knows yet whether they will work in the body.

Some diseases, such as rheumatoid arthritis and thyroiditis, are caused by an 'autoimmune' reaction – the body's defence system turns on itself. It now looks as if some of the symptoms are due to the production of cytokine, and certainly high levels of cytokines can be found in the fluid of rheumatoid joints.

The behaviour of any cell in the body depends on a number of outside signals, including cytokines. In some diseases, particularly cancer and autoimmunity, there is an imbalance in these signals. A full understanding of such imbalances could lead to highly sophisticated and specific treatments.

16 June 1988

Further Reading

Cytokines in Cancer Therapy by Frances Balkwill (Oxford University Press, 1989) provides a very detailed background to the subject. 'Tumor necrosis factor', an article by Lloyd J. Old (*Scientific American*, May 1988, p.41), concentrates on one of the cytokine family. *Reshaping Life* by G. J. V. Nossal (Cambridge University Press, 1990) is a good introduction to the main issues in genetic engineering. *Immunology* by Ivan Roitt, Jonathan Brostoff and David Male (Churchill Livingstone, 1989) is a medical textbook but nevertheless contains much of interest to the general reader.

CHAPTER 25

Cancer and Oncogenes

Richard Vile

Imagine the Earth as a single 'body', in which the people form the individual working parts. Each person would have a function that contributed both to the running of the constituent 'organs' (the countries) and to the day-to-day operation of the body as a whole. Such global cooperation would require intricate coordination between each and every individual. Detailed regulations would control each person's activities. Close supervision of reproduction would be necessary, to ensure that everyone would be replaced when they died, at the correct rate and in the correct proportions. This Orwellian picture of the world may seem far-fetched, not least because of the enormously complex organization it would require. Yet every one of the billions of cells in our bodies carries, and can read, a program of this level of complexity, which tells it when and how to develop. The result is the integrated collection of cells that cooperate to form a human body.

It is when the reading of the program breaks down that we see disease. By studying the breakdowns, scientists are beginning to understand the molecular events that give rise to many genetic diseases. Cancer is one of the main diseases that result when the population controls in the program fail, leading to uncontrolled proliferation of cells. The information for this highly complicated program is contained, in code form, in the large molecule known as **deoxyribonucleic acid**. This molecule, **DNA** for short, is found in the nuclei of cells. The DNA consists of a linear sequence of four smaller component molecules, called **bases**. The order of the bases, called A, T, C and G, forms a code which the cell copies into a messenger molecule called **ribonucleic acid**, or **RNA**. This molecule, which also consists of a sequence of bases, carries the information out of the nucleus and into the cytoplasm, where **proteins** are made. Proteins form the major components of each cell. They can form part of the structure of the cell and, among their other functions, they can also serve as 'signalling molecules' within the cell or between cells. The code contained in the RNA determines the sequence of the building blocks, called amino acids, that make a protein.

The information needed to make proteins is carried in our genes. Broadly speaking, a **gene** is a sequence of bases in the DNA that specifies the structure of a protein. When a gene makes the protein that it codes for, scientists say that the gene is **expressed**. Our genes are held in large assemblies, known as **chromosomes**, which are sometimes large enough to be seen with an optical microscope. Every human cell carries, in its nucleus, 23 pairs of chromosomes (46 in all). The exceptions are sperm and egg cells (the gametes), which contain only a single copy of each chromosome.

Single cells
Whole organism

When a sperm fuses with the egg, a single cell called the **zygote** results. The zygote carries the normal complement of 46 chromosomes, with two copies of each gene, one copy inherited from the egg and the other from the sperm. Every cell in an adult is derived from the zygote by a series of repeated cell divisions. As the body develops, the dividing cells become specialized by **differentiating** into a variety of cell types, such as muscle, liver or brain cells. Once a cell has reached a certain stage in the process of differentiation, known as **commitment**, it can no longer revert to any of the other cell types in the body. When a cell becomes committed to a certain pathway of

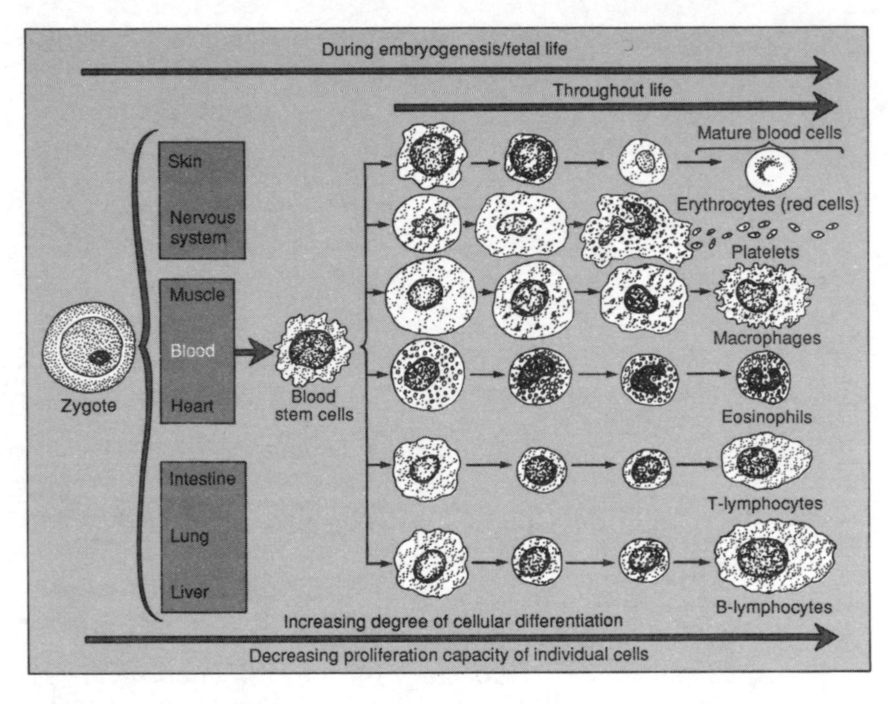

Every cell in the adult is derived from the zygote by a series of repeated cell divisions. During the development of the individual, dividing cells become specialized into a variety of cell types, such as muscle, liver or brain cells.

Scarce stem cells divide to produce two daughter cells: one remains as a stem cell; the other has the capacity to differentiate into a specific lineage. After a certain stage, a cell becomes committed to a pathway of differentiation.

For each specific type, cells become specialized for a variety of functions. Red blood cells, for example, represent a specific lineage within the blood

differentiation, one of the many specific genetic programs within its DNA becomes activated. This activation occurs in response to a wide variety of signals from the cell's environment, which are partly determined by the age and position of the cell in the developing body. The signals may be in the form of proteins sent by the cell's near or far neighbours. (These proteins are termed **growth factors** and **hormones**, respectively.)

All the cells of one specific type (those of the blood, for example) are described as belonging to the same differentiation **compartment**. Within a compartment there are a variety of cell

lineages, specialized cells with different functions. All cells of the compartment will share the same genetic program of differentiation that has been 'switched on' in the stem cells (making all the cells into blood cells, for example). The mature cells of that compartment will elaborate on the basic program to produce the specialized cells of the separate lineages. The stem cells not only express the genes required to determine the functions of the daughter cells. They also have to divide rapidly and plentifully, to ensure that each compartment has enough cells (usually many millions of cells in each lineage) to carry out its functions in the body.

So the stem cells have to express genes whose protein products drive cells through repeated divisions. These are the genes of cellular proliferation. In this way the single cell of the zygote is eventually converted into billions of cells of varying specializations. In the progression from stem cells to mature, differentiated cell types, the genes of cellular proliferation generally become less active. Simultaneously, the activity of those genes concerned with differentiation increases.

One way of defining a cancer is as a population of cells that has gone a certain distance along its pathway towards maturation, but in which the processes of proliferation and differentiation have become uncoupled. The result is that cancer cells divide rapidly and are no longer able to complete their program of differentiation. Cells with these qualities are said to be **transformed**. A population of transformed cells usually derives from the divisions of a single cell that has accumulated damaging changes in its genes. A group of similar cells (termed 'clones') derived in this way from a single parental cell is known as a **clonal population**. We do not know how often potentially cancerous cells arise in the body, or exactly what the role of the immune system is in dealing with those rogue cells. But we do know that these changes can happen to most cells, diverting them from their normal program of differentiation, within most of the compartments of the body, and at many different stages between stem cell and the fully differentiated state.

Cancer eventually kills through the damage caused by the expanding number of cancer cells. The wrong types of cell progressively occupy vital room in the body: their presence hinders normal cellular activity. Many cancer cells also eventually become able to spread from the site of origin to distant regions of the body. This process, which is called **metastasis**, is life-threatening as it increases the ability of the rapidly growing population of cells to damage the body.

Scientists name the study of cancer 'oncology'. They have known for many years that damage to a cell's DNA, known as **mutation**, is associated with the changes that lead to cancer, a process called carcinogenesis or oncogenesis. But current ideas on how genes change in cancer began to take shape only when researchers discovered small viruses called **retroviruses**. These contain up to about 10 genes, in contrast to the 10 000 or so that are present in human DNA. When a retrovirus infects a cell it copies its genetic material, which is in the form of RNA, into DNA. (This is in the opposite direction to the usual flow of genetic information, which explains the name of this group of viruses: 'retro' means backwards.) The virus then inserts its DNA into the DNA of the cell. There, hidden from the body's immune system, the viral genes can direct the cell to make viral proteins, which assemble themselves into new viruses.

In 1911 an American scientist called Francis Peyton Rous described an agent that, if passed from a chicken with a certain type of tumour to a healthy chicken, would cause a similar tumour to develop in the second bird. Later studies showed that this agent was a retrovirus, and that a specific viral gene, called **v-src**, was responsible for the tumours. Subsequent research uncovered more retroviruses, which could transform cells grown in the laboratory, making them grow rapidly and uncontrollably. These viruses also contained genes that seemed to play a part in carcinogenesis. Scientists coined the term **oncogene** to describe these genes, which may participate in converting normal cells to cancerous ones. The theory was that the retroviral oncogenes (known as **v-onc** for short) were normally dormant: only when certain agents acted on the cell did the oncogenes become active. The action of the viral proteins helped to push the cell into a proliferative (cancerous) cycle of replication.

In 1976 researchers in the US showed conclusively that sequences of DNA related to the oncogene *v-src* were present in the normal DNA of uninfected chickens. Subsequently, the rapid development of genetic engineering allowed scientists to show that other *v-onc* sequences also had their origins in the DNA of normal cells (that is, in cellular genes). In other words, some of the genes associated with the development of cancer exist within normal cells before the cells are infected by retroviruses. This discovery was so important to the current understanding of the molecular basis of cancer that it won the principal investigators, Americans Michael Bishop and Harold Varmus, the Nobel prize for medicine in 1989. Later work on the life cycles of retroviruses showed that, infrequently, they hijack (or **transduce**) incomplete portions of cellular genes into their own genetic material. This transfer damages the cellular gene so that it no longer functions correctly.

In nature, transduction of cellular genes into retroviruses is a rare event, irrelevant to the normal life cycle of the virus. But in the laboratory it has proved enormously useful in enabling scientists to identify other oncogenes. Researchers have also identified new oncogenes by a process known as DNA transfection. DNA extracted from rodent or human tumours can be transferred ('transfected') into a particular type of mouse cell, called NIH3T3, which can be grown in culture dishes. Usually, NIH3T3 cells in culture will stop growing when the cells reach a certain density – a phenomenon which is called contact inhibition. If an oncogene is introduced into these cells, they lose this restraint and grow by piling on top of each other to give a **focus** of transformed cells which researchers can detect. When scientists examine the cells from the focus they can identify the DNA sequences that originally led to the transformation of the NIH3T3 cells. Such sequences frequently turn out to contain an oncogene.

How oncogenes work
Switched-on genes

A cell receives signals that tell it to proliferate when proteins bind to **receptor molecules** (themselves proteins) located on the outer surfaces of the cell. The signal is transmitted into the cell, across the cytoplasm and into the nucleus. There, the cell responds by switching on the genes that control proliferation.

Most oncogenes appear to be altered forms of genes that code for proteins involved in the signal pathways of cellular proliferation. The normal, intact cellular genes of these pathways are known as **proto-oncogenes**. The oncogenes are often changed (in comparison with their parent proto-oncogenes) in such a way that the proliferative signals are jammed in the 'on' position. Some proto-oncogenes identified by researchers are genes containing the information for growth factors and the proteins on cells that act as receptors for growth factors. Some proto-oncogenes code for proteins that carry the signal telling the cell to proliferate across the cytoplasm into the nucleus. Still others code for proteins in the nucleus that appear to play a part in triggering DNA synthesis before cell division.

Scientists have now identified more than 30 oncogenes; some of the most closely studied are the **growth-factor receptors**. These proteins span the membrane of the cell. The part of the protein outside the cell binds the growth factor, altering the three-dimensional structure of the protein in the membrane. These changes are transmitted to the part of the protein that lies within the cell. The receptor often responds by adding a phosphate group to an amino acid called tyrosine. The tyrosine molecule is either located on the receptor, or on another molecule further along the signalling pathway. Adding the phosphate group to tyrosine is often a first step in the chain of events that eventually leads to the cell dividing.

Comparing the DNA sequences of normal proto-oncogenes with their analogous oncogenes has revealed the mechanisms by

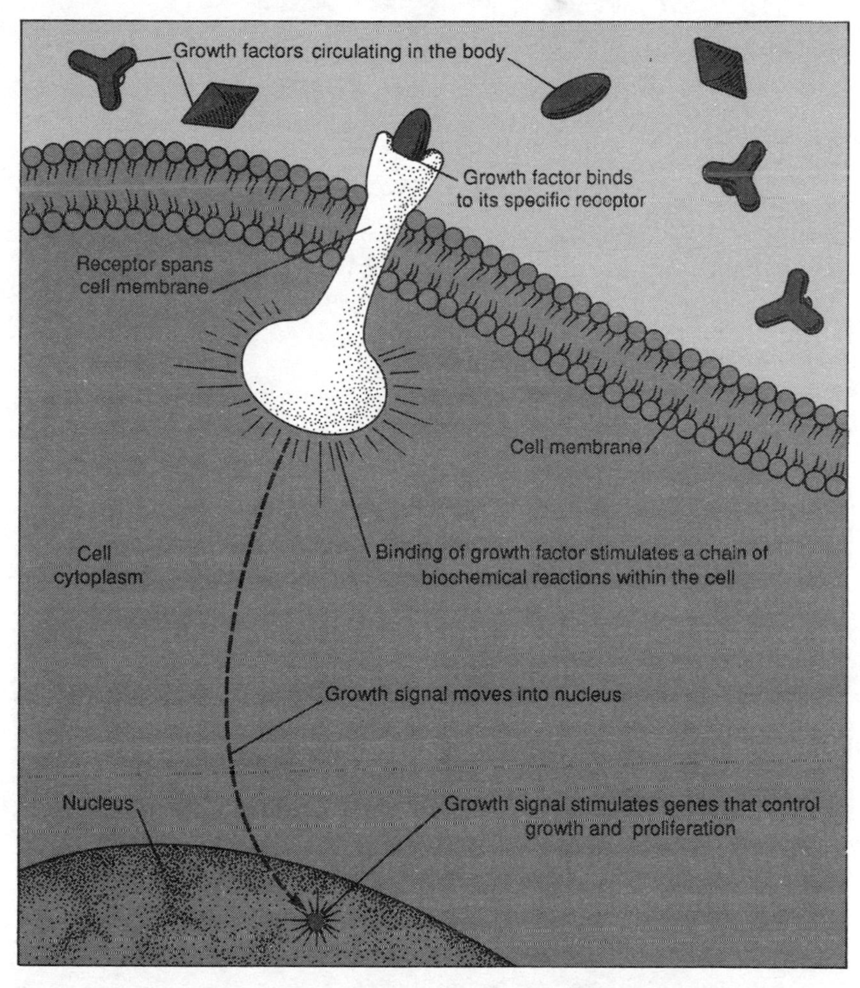

Of the many oncogenes now known, one of the most closely studied groups is the growth-factor receptors. These proteins span the membrane of the cell. The part of the protein outside the cell binds the growth factor, causing alterations to the three-dimensional structure of the protein in the membrane. These changes are transmitted to the part of the protein that lies within the cell, and from there signals pass into and through the cytoplasm and thence on to the nucleus. A cancer cell may receive continuous signals to proliferate, which mask signals telling it to differentiate or to stop growing

which proto-oncogenes become 'activated', or converted into oncogenes. The type of activation depends on the proto-oncogene concerned. Sometimes, activation results in the proto-oncogene producing too much of its protein product at inappropriate times. Here, the protein product of the proto-oncogene is not structurally altered and activation of its oncogenic potential is the result of abnormal regulation of its expression.

As a result of such activation, cells may receive continuous signals to proliferate, masking those telling them to differentiate or to stop growing. Some viruses may influence the expression of a proto-oncogene in this way when they insert their DNA into the

If the cell's chromosomes break and rejoin with the wrong partners, a process called chromosome translocation, two mechanisms are possible to create an oncogene. The first leaves the structure of the protein intact but disrupts the regions controlling its levels of expression. The second alters the actual structure of the protein

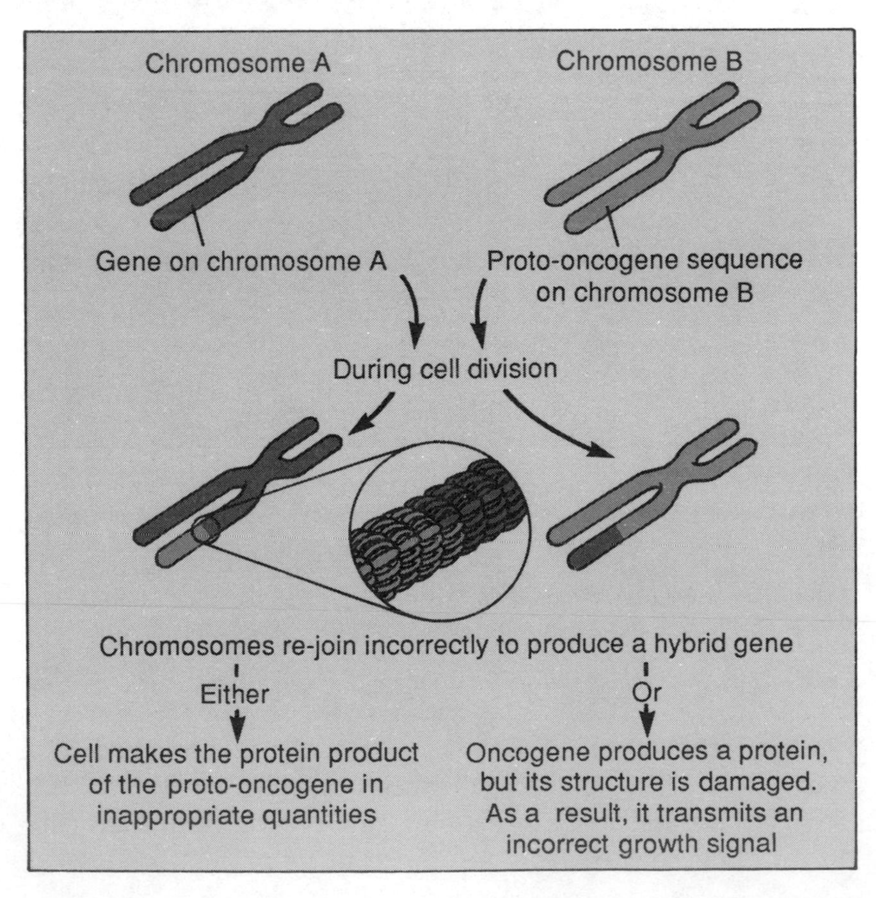

DNA of the cell. Scientists call this effect 'insertional mutagenesis'. Such an effect may also be apparent if the cell's chromosomes break and rejoin with the wrong partners, a process called 'chromosome translocation'. For example, in the human cancer called Burkitt's lymphoma a gene involved in DNA synthesis, the *myc* gene, becomes linked to the wrong region of another chromosome. As a result it is permanently expressed even in the face of signals that tell the cell to slow its rate of growth. In other human tumours there are many copies of certain proto-oncogenes instead of just the normal two copies. Scientists say that these proto-oncogenes are 'amplified'; as they contribute to the development of the cancer they too are called oncogenes.

Proto-oncogenes may become activated when their genetic structure is altered. Such changes may occur as a result of chemical or physical carcinogens, such as ionizing radiation. Other causes of mutation include mistakes made by the cell's machinery for replicating DNA, chromosomal translocations (as described above), or infection by retroviruses. When the genetic sequence of a proto-oncogene changes, the protein that the gene makes may be damaged. One class of proto-oncogenes has products called the **ras proteins**, which are involved in signalling. Changing a single base in the sequence of the *ras* gene, thus altering a single amino acid in the protein, is enough to ensure that the signalling activity of the *ras* proteins is continually turned on. Up to 15 per cent of human cancers have a mutation of a *ras* gene of this kind, although their exact role in the development of cancer is not clear.

Other changes can include large-scale alterations in the structure of the protein product of the proto-oncogene. For example, one proto-oncogene codes for the protein receptor of a substance called epidermal growth factor. The proto-oncogene can become converted to the oncogene known as *erb B*. This conversion involves the loss of the section coding for the part of the receptor normally found outside the cell. Researchers believe that, as a result, the receptor continually transmits a growth signal.

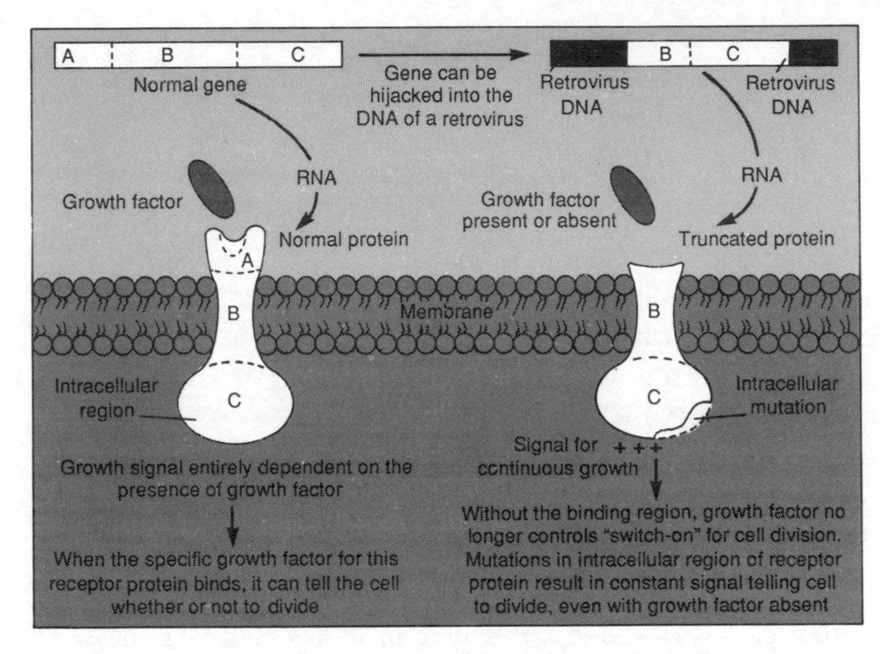

Large-scale alterations in the structure of the protein product of the proto-oncogene may play a part in the development of a cancer. For example, several proto-oncogenes code for the protein receptors for growth factors. In this example, the proto-oncogene can convert to the oncogene. Such a conversion may involve loss of the section coding for the part of the receptor outside the cell. This loss may cause the receptor to transmit a continuous growth signal

Other influences
Multiple causes

Rarely, if ever, is it possible to attribute the metamorphosis of a normal cell to a cancer cell solely to the activation of a single proto-oncogene. The genes in the cell from which the cancer population originates usually undergo many different changes; each change is alone not sufficient to transform the cell, but the accumulated changes lead to a cancer. Activation of proto-oncogenes is just one stage in this process. Transformed cells usually express at least two oncogenes that work together, as well as showing other genetic abnormalities.

A range of genetic changes dictates the behaviour of any cancer. These changes include alterations in genes that control:

a) how rapidly and extensively the cancer will spread through the body

b) whether or not the cancer cells can escape the body's own natural mechanisms aimed at killing them

c) the development of resistance to the drugs that doctors normally use to kill rapidly dividing cells.

For many years, scientists believed that the genetic mutations responsible for cancer deleted genes that restrained cell growth. The discovery of oncogenes led to a new perspective of cancer as a disease caused by genes that actively promote uncontrolled growth. Recent discoveries have begun to draw these two notions together, allowing researchers to produce a unifying concept of the changes that occur in the cancer cell.

The study of some rare inherited human cancers has suggested that there exist specific genes whose products restrict the proliferation of cells, in contrast to the products of the oncogenes. These signals tell the cell to slow down its growth and to increase its level of differentiation (see figure on page 316). The loss of both copies of these genes causes the cell to 'miss' the crucial signpost that would lead it from the proliferative pathway and towards differentiation. Once the cell has taken the wrong 'path' it cannot retreat from its program of cell divisions; subsequent activation of oncogenes will push such a cell into uncontrolled proliferation. The genes that exert this negative control are called anti-oncogenes or tumour suppressor genes, because they appear to be necessary to allow normal cells to differentiate into mature cell types that have little or no potential for growth. Such genes reside at sites where, in some cancer patients, segments of chromosomes are missing. Scientists have recently identified the first of this class of genes by studying children with a rare eye tumour called retinoblastoma. Both copies of the gene, called the *Rb* gene, were absent or mutated in these children. Whereas the presence of oncogenes is required to promote the development of cancer, it is the absence of tumour suppressor genes that seems to be necessary in order to lift the normal controls on growth and differentiation of cells.

The ultimate goal of research into the genes that contribute to the cause of cancer is to obtain better treatments for the disease – although many years of research into oncogenes have still not made a major impact on the therapies that doctors use. There is much hope for the future, however. Knowledge of the oncogenes active in a given cancer can sometimes be helpful in predicting the course of the disease and its outcome, so helping the doctor to gauge the timing of treatment. It may also be possible to design drugs to act on cells that are expressing the protein product of a known oncogene; it may even be feasible to exploit the differences between the normal protein and the mutant product to kill the cancer cells. The discovery of tumour suppressor genes may provide the opportunity to predict which people are most at risk from certain cancers; individuals without certain genes may be predisposed to specific cancers. Finally, there are hopes, albeit remote, that replacing the missing functions of a tumour suppressor gene may prove easier than correcting the aberrant behaviour of an oncogene. Gene therapy for cancer may yet be a possibility.

10 March 1990

Further Reading

Molecular Biology of the Cell edited by Bruce Alberts and others (Garland, 1983). *Molecular Biology of the Gene* edited by J. D. Watson, N. H. Hopkins and others (fourth edition, Benjamin-Cummins, 1987). *The New Genetics and Clinical Practice* by D. J. Weatherall (second edition, Oxford University Press, 1990). *The Molecular Basis of Cancer* edited by P. B. Farmer and J. M. Walker (Routledge, 1985) is slightly more complex.

PART 4

Some Chancy Mathematics

CHAPTER 26

Risky Business

Ian Stewart

Thousands of people died in the Armenian earthquake of 1988. We shall probably never know the precise total. Nearly 200 people died in the San Francisco quake in 1989. When the *Herald of Free Enterprise* turned over in the cold waters outside Zeebrugge in 1987, the death toll eventually reached 193. Less headline-catching is the routine death of 15 people, on average, every day on Britain's roads. All human actions – including inaction – carry an element of risk.

How much risk depends on the activity. Leaping across a canyon on a motorcycle carries a large degree of risk; watching television in the living room a small one. Politicians may say that the risk of a particular accident is one in a million. To understand this statement you need a clear grasp of the scope and the limitations of the mathematics that underlie it. You also need to know what assumptions it is based upon.

Taking a chance
The meaning of risk

The degree of risk is the extent to which harm may result from a particular action. There are many different kinds of risk, including economic risk, the risk to life and health and the risk to the environment. Risk is not the same as danger. Walking downstairs carries a definite degree of risk: it is one of the commonest causes of accidents in the home. But you would hardly call it dangerous. In our daily lives we all risk being involved in accidents. If you ride in a car, it may crash. If you

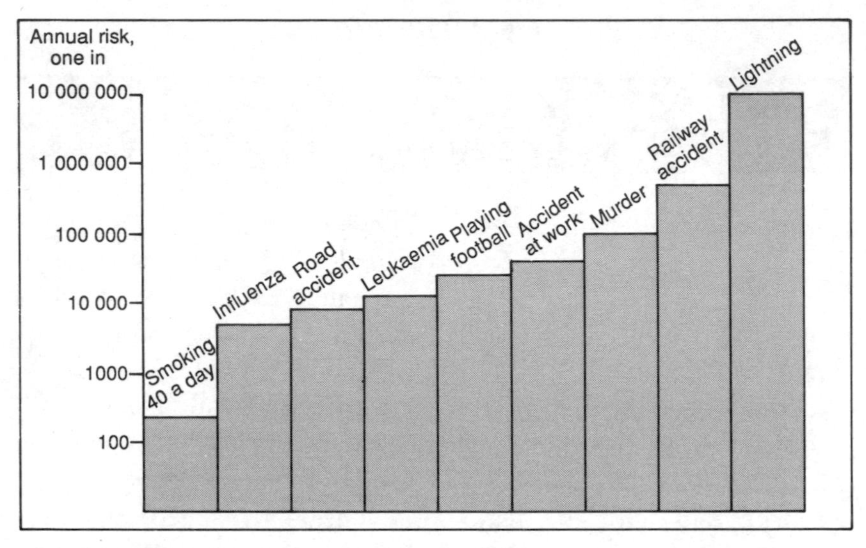

The risk of dying from some of the more common, or noteworthy, causes

choose to walk, a car may hit you. If you stay at home and light the gas, the house may catch fire.

Medical science has reduced the risk of catching a serious disease, but it has not eliminated it – people still suffer from AIDS, legionnaires' disease and food poisoning. Because it is impossible to eliminate risk, the best that we can do is to balance risk against benefit. Far more people would die of cold if the government banned gas fires than die in house fires or gas explosions because of them. So the benefit greatly outweighs the risk, and the decision is easy. Nuclear power is less straightforward. There are benefits. Nuclear power produces less acid rain than does the process of making electricity from burning fossil fuels. Producing nuclear electricity does not involve the death of a single coalminer. There are risks, however, such as pollution or a catastrophic release of radiation in a major accident, as well as deaths among uranium miners. The mathematical analysis of risk provides objective and rational methods for making such judgements. But just because a method is objective and rational, that does not mean it is necessarily right. The mathematics depends upon assumptions

Probability rules

Rule 1: If the probability of an event A is $p(A)$, then the probability that A does not happen is:

$$p(\text{not } A) = 1 - p(A)$$

Rule 2: Two events are mutually exclusive if they cannot both occur together. For example, if you throw a die, then the events 'a six' and 'a five' are mutually exclusive, because you cannot throw both at once. The probability that one or the other occurs is the sum of the separate probabilities:

$$p(A \text{ or } B) = p(A) + p(B)$$

Rule 3: Suppose two experiments occur in succession. They are independent if the outcome of the first has no effect on the outcome of the second, such as tossing a coin and then drawing a card from a pack. What happens to the coin does not affect the drawing of a card from the pack. You can find out the probability that both will occur by multiplying the separate probabilities:

$$p(A \text{ and } B) = p(A)\, p(B)$$

Rule 4: This is an approximation that mathematicians often use in risk calculations. Suppose that A and B are events, independent but not necessarily mutually exclusive, whose probabilities $p(A) = a$ and $p(B) = b$ are very small. What is the probability that at least one of them happens? It should be the sum of the following probabilities:

$$p(A \text{ and not } B) = a(1-b)$$
$$p(\text{not } A \text{ and } B) = (1-a)b$$
$$p(A \text{ and } B) = ab$$

Thus:

$$p(A \text{ or } B) = a(1-b) + (1-a)b + ab = a + b - ab$$

If a and b are small then ab is very small. So we can neglect the term $-ab$, and so:

$$p(A \text{ or } B) = a + b = p(A) + p(B)$$

In other words, events of small probability can be considered as being mutually exclusive, even if strictly speaking they are not independent of each other.

about the real world, about how people behave, and about how accurate information is. The mathematics of risk is an aid to judgement, not a substitute for it.

Probability is a way of expressing risk mathematically. It is always a number between 0 and 1. An impossible event has a probability of 0; an event that is certain to happen has a probability of 1. Everything else lies somewhere in between. Probability is the proportion of cases in which an event occurs. For example, if you toss a fair coin, you expect to get heads about as often as tails. In the long run, about half the tosses will be heads. We say that the probability of the event 'heads' is a half. The probability of rolling a six with a fair die is one in six. We can write these either as fractions (1/2, 1/6) or as decimals (0.5, 0.167).

According to the old Central Electricity Generating Board the probability of a core meltdown in a nuclear power station was one every 10 000 years. In such an accident the uranium fuel melts along with much of the reactor, but radiation need not escape. A probability of one every 10 000 years sounds very reassuring, but it is worth taking a closer look. What it means is that for each nuclear reactor, the probability of core meltdown in any given year is one in 10 000; that is, 0.0001 per year. There are roughly 40 nuclear power stations in Britain, so the probability that at least one will have a meltdown in any given year is the sum of the 40 probabilities, which is 0.004. The probability of at least one meltdown in Britain during the next 25 years is 25 times this, or 0.1. That is, the chances are one in 10. This does not sound as reassuring as 'one every 10 000 years'. But it is just a different way of saying the same thing.

Great expectations
Pools losers

Whether or not a risk is acceptable depends not only on the probability of a harmful event but also on how harmful it is. Few people would worry if they were told that with probability

0.5 they might lose 10p. They would be less happy if the sum involved was £1 million. The simplest way to deal with the scale of loss is to multiply the probability of the event by the amount that could be lost. This is called the **expected loss**. In these two examples the expected losses are 5p and £500 000.

Cost-benefit analysis, or **risk-benefit analysis**, is a way of balancing the expected loss against the risk. In everyday life, people often take decisions where the expected benefit is negative. On average, people who take out life insurance policies or play the football pools will lose. To insure a life, people pay a certain amount of money, the premium, to the insurance company. In return, the company guarantees to pay a much larger amount when the insured person dies. The insurance companies have to make a profit to stay in business, so they employ actuaries (statisticians who specialize in calculating risks) to set the premium high enough to guarantee that on average the companies win. This means that each person they insure has an expected loss. Despite this, it makes sense to take out life insurance. The cost of the premium is relatively small, something that the person involved can easily afford to pay. The benefit if that person dies unexpectedly is large: the person's family will still have a house to live in and enough money to buy food and clothing. A similar analysis applies to football pools, and bookmakers. On average, only the pools company gains. But individuals can risk tiny amounts and have a small chance that a big win will transform their lives. The main difference between the pools and the insurance is that in life insurance the aim is *not* to win the jackpot.

One problem with 'expected value' is that it treats everything alike in all circumstances. It considers a loss of £1 to a pauper to be the same as a loss of £1 to a millionaire. We can measure things differently. If instead of the average gain or loss we consider the maximum gain or loss, then the picture for life insurance changes dramatically. For a premium of £10 per month over a period of five years, you can buy insurance cover of £20 000. The maximum loss is £600 (£10 × 60 months); the maximum gain is £20 000, less any premiums you might have paid. Of course, it is no more clear that the maximum, rather than the average, is the 'sensible' measure. Mathematics can provide a limitless range of such measures. But mathematics alone cannot tell us which, if any, is the 'right' one to use. There may be no right measure. In fact, one of the difficulties

The great railway disaster

Disasters very rarely have a single cause. Risk involves chains of cause and effect in which series of individual events combine to produce a disaster. In order to calculate the combined risk it is important to have accurate estimates of the probabilities of the individual events. One technique that is used widely in risk assessment is to construct a fault tree. This is a diagram that shows the possible chains of events leading to a harmful outcome.

A simple example is parachuting. Inside each parachuter's pack is a main parachute and a reserve. The jump is fatal if both parachutes fail to open. So the fault tree is a two-link chain. If the probability of failure is one in a thousand for each parachute then the total probability is one in a million.

For a slightly more complicated example, consider a train passing a signal. An accident may happen if either the signal fails

in the debate over nuclear power may be just this point. No matter how small the risk of a really serious accident, the effects if one were to occur would be horrifying. Is it possible to do a calculation that will balance any gain against the tiny chance of killing a million people?

---

or the driver ignores (or misreads) it. But that fault alone will not cause an accident: there has to be an obstacle on the line, such as another train. If the obstacle is close, then the train will hit it. If it is further away, the train may have time to stop – unless the brakes fail. With some rather arbitrary figures for probabilities, the fault tree looks like the diagram.

For the train we must trace all possible chains through the tree and work out their probabilities by multiplying, and adding the results together.

Two chains of causality contribute almost all of the probability here: signal failure/obstacle on the line; or, driver error/obstacle on the line. So for this particular fault tree, brake failure has little significant effect.

These calculations contain several hidden assumptions. The main one is that each branch of the tree involves mutually exclusive events which are independent of each other.

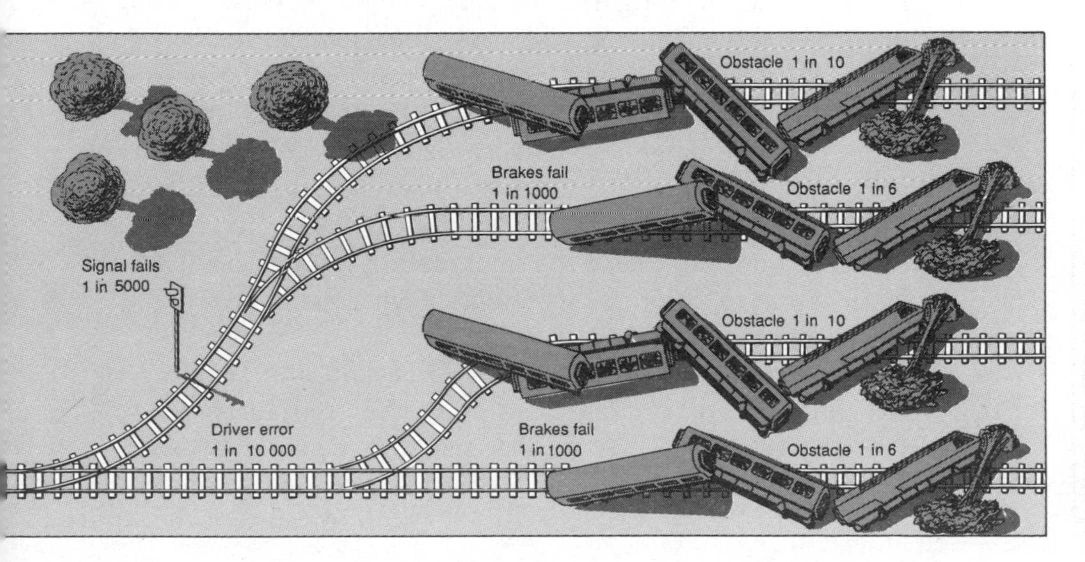

Quake hits Manchester
Calculating the odds

Every day, commercial aircraft make an enormous number of flights; every year a few of them crash. We can estimate the probability of a crash by dividing the number of crashes by the total number of flights. The more often an event occurs, the more accurately we can estimate its probability.

It is more difficult to estimate the probability of a rare event in this way. For example, what is the probability of a *severe* earthquake happening in the Manchester area? No one has ever recorded such an event, so we might estimate the probability as 0. But this must be an underestimate. Even though earthquakes are rarer in Britain than they are in other parts of the world, such as California or Japan, they do happen. In April 1990 the second largest earthquake in Britain this century, registering 5.2 on the Richter scale, hit Clun, in Shropshire. The probability of a big earthquake in Manchester is very low, but it is extremely difficult to say how low.

Unexpected sources of risk cause even more problems. Before manufacturers began to use chlorofluorocarbons (CFCs) in aerosols, they investigated the likely effect of these chemicals on the environment, including possible damage to the ozone layer. The researchers chose CFCs because they are unusually stable compounds, and so are unlikely to react with atmospheric ozone. Unfortunately, nobody realized that ice crystals in the upper atmosphere would make their reaction with ozone much more likely. If your analysis of risk omits a major hazard because you do not have the imagination to consider it, or the information to work out its effects, then your analysis may be misleading.

The way in which researchers collect data and analyse them is also important. Averages can conceal more than they reveal. Britain has occasional high winds. When the wind gusts above 60 knots it can damage buildings. The average wind speed is less than 10 knots. However, this does not mean that the wind will never cause structural damage. In three out of the past 30 years, wind speeds at Heathrow airport have exceeded 60 knots, including during the great storm of 1987 which damaged buildings and uprooted trees over a swathe of southern Britain. Another example is assessing the rate at which AIDS is transmitted. Several traditional methods work with averaged figures for infection, rate of sexual contact, and so on. But the risks differ widely for different groups, because they depend on lifestyles. Using averages in these circumstances may produce nonsensical figures.

Another problem is finding out whether a hazard really is the cause of some adverse event. Ten years ago, people argued fiercely about whether cigarettes cause lung cancer (some tobacco companies still deny that this is true). As recently as six years ago the British government was disputing that lead in petrol is dangerous, even though lead is known to impair children's intelligence. A topical example of this problem is leukaemia clusters. Leukaemia is a fatal cancerous disease of

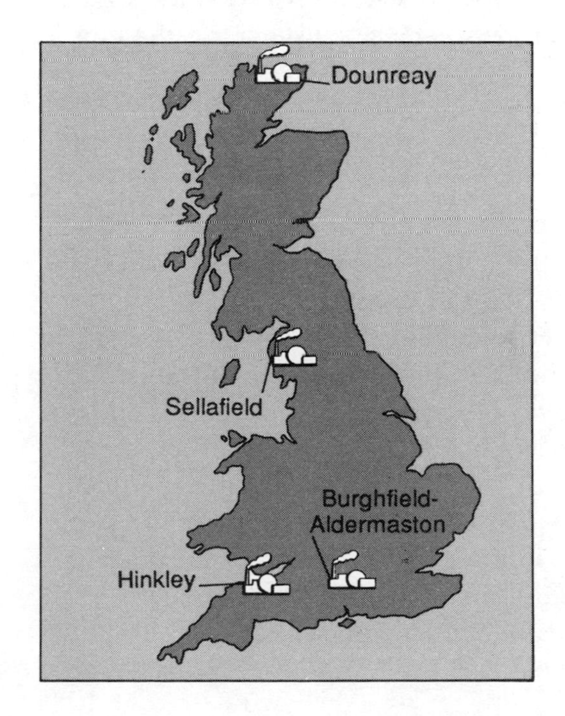

The four British nuclear sites with leukaemia clusters. Scientists have not discovered clusters at 20 other sites

the blood, which affects children in particular. In some parts of Britain there is an unusually large number of leukaemia cases. Researchers have discovered some of these 'clusters' near nuclear power stations and related nuclear installations, such as Sellafield. Does this mean that nuclear power stations cause leukaemia? Until recently that was unclear. First, some clusters are well away from nuclear power stations. Secondly, some nuclear power stations do not have clusters nearby. And as the number of leukaemia cases involved is relatively small, chance effects may also be operating. This makes it extremely hard for researchers to decide what is really going on. In February 1990, however, Martin Gardner, of the University of Southampton, concluded that there was a link. He said that the exposure to radiation of men working at Sellafield appears to have caused genetic damage in children they fathered. British Nuclear Fuels at first advised workers exposed to the highest radiation not to have children, although it has since withdrawn this advice. In this case the link between radiation and children seems to be through the father, which explains why previous studies failed to find a direct connection between the clusters and the power stations.

This example shows how important it is to distinguish between **correlation** and **causality**. Two events are correlated if one is often accompanied by the other. To find a correlation is easy: you just see how often the two events occur together. Causality is one event causing another. To prove causality you must pin down the entire chain of events, and that can be very difficult. Companies often complain that evidence that their product is unsafe 'does not establish causality'. Although correlations may not prove causality, they certainly do not disprove it. In fact, all scientific evidence about causes is really about correlations. We observe in hundreds of experiments that one event leads to another. For example, litmus paper turns red when we put it in acid, and we assume that putting it in acid is what turns the litmus red. This is just a very high degree of correlation; it could all be coincidence. In the end we have to use some common sense.

Acceptable risk
A sporting chance

Someone must decide at some stage whether or not a risk is 'acceptable'. But what is an acceptable risk?

There is a bad answer: a risk is acceptable if we get the benefit while others suffer the effects. Dumping my toxic waste is fine provided you keep it well away from me. A drug is 'safe' if I make the profits by selling it and other people run the risk of using it. All too often, decisions may be biased – perhaps unconsciously – by this kind of reasoning.

A better answer is that the benefits must outweigh the risks for most of the people involved. Most people think that the convenience of car travel outweighs the very distinct risk of being involved in a serious accident. People who indulge in 'dangerous' sports, such as mountaineering, consider that the fun outweighs the risk – or perhaps they underestimate the risk.

The way people react to a risk does not always reflect its probability. For example, the probability of being killed in a terrorist attack on an aircraft is smaller than the probability of dying because the shuttle bus crashes on the way to the airport, but most travellers are more worried about the terrorist attack. The probability of getting rabies, even in countries where it is common, is very low – a few cases per million people per year. Yet the British government devotes much effort and money to very strict quarantine regulations to keep rabies out. Is this sensible, or a waste of money?

Mathematics does not answer this question. It can give us a good idea of the dangers involved in some activity, and mathematical assessments of levels of risk provide useful information for public debate. But the mere fact that a mathematical calculation produces a very small figure for the level of risk does not mean that the risk is automatically 'acceptable'. For example, asbestos fibre can cause fatal lung disease. The probability of contracting such a disease is very low. But that does not mean we should go on using asbestos for ceiling tiles or brake linings. In any case, very low may not be as low as it sounds: in a population of 50 million, a disease with an annual probability of one in a million will kill 50 people every year. Again, we have to use our judgement and common sense: the numbers alone cannot take the decisions for us.

Before the Challenger disaster, in 1986, the officially estimated figure for an accident involving a space shuttle was one in 100 000. This calculation was ridiculously optimistic. It failed to take account of the way events might depend on one another. In particular, launching a shuttle on a cold day substantially increased the probability of several different events. In the subsequent inquiry, the American physicist Richard Feynman made the point that the calculations looked very fishy. An engineer eventually said the true figure was around one in 300.

Another problem with many methods of risk assessment is that inaccuracies in the figures can change the result of the calculation. This is important. In practice, the estimate for a single event can easily be 10 times too large or 10 times too small, and the final answer may vary wildly. One method for studying the effects of possible errors is called **sensitivity analysis**. It aims to determine the range of probabilities that a calculation of risk may yield.

Human psychology
Fear and dread

The way that human beings react to risk is influenced by psychological factors. 'Familiarity breeds contempt', the old saying goes, and it might. Workers in hazardous professions often fail to take precautions. They are so used to the danger that they no longer fear it. In much the same way we fear the unknown. We often think unfamiliar hazards are more risky than they really are. Even taking precautions may not always

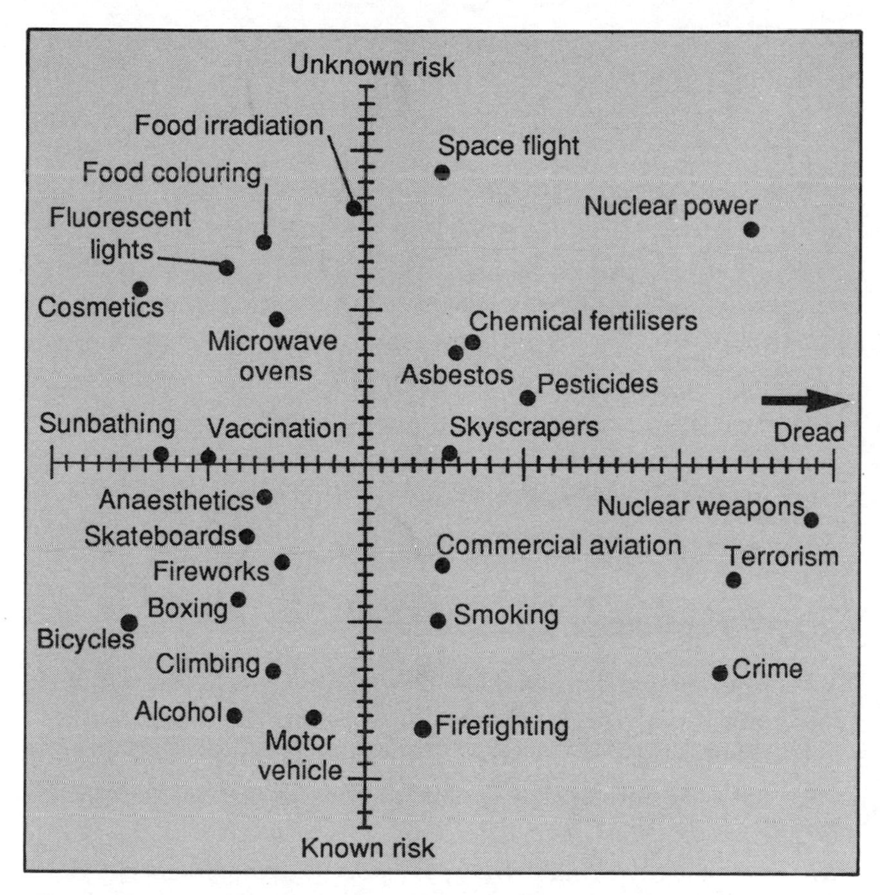

The graph plots estimates of people's ignorance of risks against their dread of those risks (taken from the Royal Society report Risk Assessment*). For example, the plot shows that people tend to dread nuclear power to the same extent as crime, yet their knowledge of the risks of nuclear power is much less than it is of crime*

be as beneficial as we think. One theory, known as the **risk compensation hypothesis**, says that safety precautions may increase the exposure to risk. For example, a trapeze artist may take greater risks when performing with a safety net than without one. When drivers wear seat belts they may feel safer, hence drive less carefully: the end result could be more accidents rather than fewer.

People often assess probabilities wrongly. A common

example is the **law of averages**, according to which lightning never strikes in the same place twice. In war, soldiers often shelter in shell craters on the grounds that a hit by a second shell is less likely. Neither of these beliefs is true. Lightning is more likely to strike the same place again, because it tends to favour exposed places on high ground. Artillery shells do not remember where previous shells have gone. (Though sheltering in a shell crater is perfectly sensible, because you can keep your head below ground level.) Perhaps the best exposure of the 'law of averages' is the story of the man who always took a bomb with him when he travelled by plane, on the grounds that two bombs on the same plane is such an improbable occurrence that in practice it could never happen.

19 May 1990

Further Reading

Risk and Chance edited by Jack Dowie and Paul Lefrere (Open University Press, 1980). *Acceptable Risk* by Baruch Fischhoff, Sarah Lichtenstein, Paul Slovic, Stephen L. Derby and Ralph L. Keeney (Cambridge University Press, 1981). *How to Tell the Liars from the Statisticians* by Robert Hooke (Dekker, 1983). *Risk Assessment: a Study Group Report* (Royal Society, 1983). *Lady Luck* by Warren Weaver (Dover, 1963). *What do You Care What Other People Think?* by Richard P. Feynman (Unwin, 1988). *The BMA Guide to Living with Risk* (Penguin, 1990).

Index

Page numbers in *italic* type refer to illustrations.

READ MORE IN PENGUIN

In every corner of the world, on every subject under the sun, Penguin represents quality and variety – the very best in publishing today.

For complete information about books available from Penguin – including Puffins, Penguin Classics and Arkana – and how to order them, write to us at the appropriate address below. Please note that for copyright reasons the selection of books varies from country to country.

In the United Kingdom: Please write to *Dept. JC, Penguin Books Ltd, FREEPOST, West Drayton, Middlesex UB7 OBR*

If you have any difficulty in obtaining a title, please send your order with the correct money, plus ten per cent for postage and packaging, to *PO Box No. 11, West Drayton, Middlesex UB7 OBR*

In the United States: Please write to *Penguin USA Inc., 375 Hudson Street, New York, NY 10014*

In Canada: Please write to *Penguin Books Canada Ltd, 10 Alcorn Avenue, Suite 300, Toronto, Ontario M4V 3B2*

In Australia: Please write to *Penguin Books Australia Ltd, 487 Maroondah Highway, Ringwood, Victoria 3134*

In New Zealand: Please write to *Penguin Books (NZ) Ltd,182–190 Wairau Road, Private Bag, Takapuna, Auckland 9*

In India: Please write to *Penguin Books India Pvt Ltd, 706 Eros Apartments, 56 Nehru Place, New Delhi 110 019*

In the Netherlands: Please write to *Penguin Books Netherlands B.V., Keizersgracht 231 NL–1016 DV Amsterdam*

In Germany: Please write to *Penguin Books Deutschland GmbH, Friedrichstrasse 10–12, W–6000 Frankfurt/Main 1*

In Spain: Please write to *Penguin Books S. A., C. San Bernardo 117–6° E–28015 Madrid*

In Italy: Please write to *Penguin Italia s.r.l., Via Felice Casati 20, I–20124 Milano*

In France: Please write to *Penguin France S. A., 17 rue Lejeune, F–31000 Toulouse*

In Japan: Please write to *Penguin Books Japan, Ishikiribashi Building, 2–5–4, Suido, Tokyo 112*

In Greece: Please write to *Penguin Hellas Ltd, Dimocritou 3, GR–106 71 Athens*

In South Africa: Please write to *Longman Penguin Southern Africa (Pty) Ltd, Private Bag X08, Bertsham 2013*